AGRO-ECOLOGICAL LAND RESOURCES ASSESSMENT FOR AGRICULTURAL DEVELOPMENT PLANNING

A CASE STUDY OF KENYA

RESOURCES DATA BASE AND LAND PRODUCTIVITY

Main Report

A.H. Kassam, H.T. van Velthuizen, G.W. Fischer and M.M. Shah

Land and Water Development Division
Food and Agriculture Organization of the United Nations
and
International Institute for Applied Systems Analysis

1991

Any part of the national land resources data base and the productivity models described in this report may be modified in the light of new knowledge and/or new objectives. The data base and the models are part of a larger district and national level planning tool and they are expected to be expanded and refined with use.

The designations employed and the presentation of the material in this document do not imply the expression of any opinion whatsoever on the part of FAO or IIASA concerning the legal or constitutional status of any sea area or concerning the delineation of frontiers.

M-51
ISBN 92-5-103303-X

Contents

Page

Figures

Tables

Page

APPENDIX TABLES

REPORT AND TECHNICAL ANNEXES

The work is recorded in a main report and technical annexes.

Main Report:

Agro-ecological Land Resources Assessment for Agricultural Development Planning - A Case Study of Kenya

Resources Database and Land Productivity

Technical Annexes:

1. Land Resources
2. Soil Erosion and Productivity
3. Agro-climatic and Agro-edaphic Suitabilities for Barley, Oat, Cowpea, Green gram and Pigeonpea
4. Crop Productivity
5. Livestock Productivity
6. Fuelwood Productivity
7. Systems Documentation Guide to Computer Programs for Land Productivity Assessments
8. Crop Productivity Assessment: Results at District Level

Acknowledgements

The ecological potential of land and water resources for food production and agricultural development and the appropriate policies for their management were the focal point of the FAO/UNFPA/IIASA study "Land Resources for Populations of the Future". The results of this regional agroecological zones (AEZ) study for developing countries in Africa, Southeast Asia, Southwest Asia and Central and South America were presented at the FAO Conference in 1983. The Conference recommended that the method, with a refinement of the resource data base, should be extended to detailed country planning of agricultural development. Kenya was selected for a case study to amplify and develop the methodology.

This main report presents the results of the methodological developments and the resource data base compiled for this first detailed country study, undertaken by FAO regular programme and in collaboration with IIASA and the Government of Kenya.

This work is the result of cooperation between numerous experts in different disciplines. Besides the authors, the reports have benefitted from comments and inputs from FAO staff, G.M. Higgins, R. Brinkman, M.F. Purnell, J. Antoine, F. Nachtergaele, I. Harder, C. Smith-Redfern and D. Mazzei in the Land and Water Development Division; A.W. Qureshi, R. Sansoucy and P. Hassoun in Animal Production and Health Division; E Wagner, F. Riveros and J.M. Suttie of Plant Production and Protection Division, A. Christoforides and G. Pietrantoni in Computer Service Centre; B.P. Dutia, M.H. Abbas and A. Zichy in Economic and Social Policy Department; C.H Murray and K.D. Singh in Forestry Department, and many others.

The cooperation with the Government of Kenya in initiating this case study has been decisive for its implementation. Invaluable early support and assistance was provided by Professor P. Ryan and members of his staff in the Ministry of Economic Planning. Many other Government Departments and Agencies in Kenya provided important insights and made available their data and information. Their contributions and assistance are gratefully acknowledged.

Finally, the support of the Managements of FAO and IIASA, and the Government of Kenya have made possible the completion of this work.

Chapter 1

Introduction

The ability of land to produce is limited and the limits to production are set by climate, soil and landform conditions, and the use and management applied to the land. Accordingly, knowledge on land resource endowment and its potential is an essential prerequisite to planning of optimum land use and subsequent sound 'long-term ' agricultural and economic development.

In particular, for planning optimum land use, answers are needed to the following questions:

- is there sufficient land to meet future food and agricultural needs?
- where are the potentially utilizable areas and what are their extents?
- for which land uses are they suitable and what is the range of their potential?
- which level of technology is required under these various circumstances?
- what is the risk of degradation and what measures are required to minimize the risk?
- where can maximum returns from increased inputs be obtained and on what land uses?
- what levels of investment are needed to obtain these returns?
- what are the limitations to production increases?
- where should research, extension and education efforts be concentrated?

Aware of these facts FAO began in 1976 the Agro-ecological Zones Project (AEZ, FAO 1978-81) to assess production potential of land resources in the developing world, and to provide the physical data base necessary for planning future agricultural development. Climate, soil and landform data were combined into a 1 : 5 million scale land resources data base of about 91 000 unique agro-ecological cells. For each of these, crop requirements and crop growth models were applied to estimate rainfed yields and output at a range of levels of agricultural inputs.

This subsequently made it possible for FAO to undertake, with support from UNFPA, and in collaboration with IIASA, assessments of the potential population supporting capacities of 117 developing nations, grouped into five regions - Africa, South-west Asia, South-east Asia, Central and South America (FAO 1980, 1982). The methodology and the findings were discussed at the 1983 FAO Conference which, recognizing the importance of such work for development, recommended that future activities be concentrated at the national level (FAO 1984a).

The regional assessments, in effect, ascertained country situations within a regional context; the national assessments of land productivity and population supporting capacity are intended to quantify sub-national situations within national contexts.

The programme of work covered under the title: 'Agro-ecological Land Resources Assessment for Agricultural Development Planning, a Case Study of Kenya ' is concerned with the development and implementation of a national level methodology for the determination of land use potentials of land resources in each of the 41 districts in Kenya, as a tool in policy formulation and development planning. This case study has been carried out by FAO and IIASA in collaboration with the Government of Kenya (FAO 1984b), and is part of the follow-up programme thrust to implement the 1983 FAO Conference recommendations.

The national methodology developed for Kenya is based on the principles fundamental to any sound evaluation of land. These principles are described in 'A Framework for Land Evaluation' (FAO 1976). The set of methods used in the compilation of the national land inventory as well as the land use productivity models, have followed the philosophy and concepts developed in the FAO-AEZ regional assessments of land and population potentials of land resources of individual nations (FAO 1978-81, 1982).

The main objective of the Kenya national assessment is to apply the methodology to quantify:

(i) how much land of different quality is contained by each district in Kenya;

(ii) what alternative kinds of land uses can be considered on land of different qualities in different districts, and what are their productivity potentials at different levels of production inputs;

(iii) how many people can be supported at those different levels of production inputs, and at what costs; and

(iv) what are the policy implications of these land and population potentials for food and economic self-sufficiency, when examined against the background of present and future populations, food and agricultural demands, and socio-economic needs, opportunities and constraints.

All four above mentioned objectives are addressed through methods and techniques that can operate within Kenya for future use (and refinements) by national planners and experts.

This Main Report presents: (a) the 1:1 million scale computerized land resources data base of Kenya, and (b) the crop, livestock and fuelwood productivity models developed for the estimation of potential productivity of land resources. An overview of the methodology is presented in Chapter 2. The land resources data base is presented in Chapter 3, which is followed in Chapter 4 by the model to assess soil erosion hazard and its impact on productivity. Chapter 5 presents the crop productivity model, which is followed in Chapter 6 by the livestock productivity model. The report ends with the fuelwood productivity model which is presented in Chapter 7. The main report is supported by technical annexes which deal with details.

The technical annexes are:

1 Land Resources
2 Soil Erosion and Productivity

Chapter 2

Methodology

2.1 Procedures

The overall methodology is schematically presented in Figure 2.1, and comprises the following activities:

(i) formulation and selection of crop, livestock and fuelwood land utilization types (LUTs);

(ii) determination of ecological requirements of crop, livestock and fuelwood land utilization types;

(iii) compilation of climatic resources inventory;

(iv) compilation of soil and landform resources inventory;

(v) compilation of land use inventory (including socio-economic aspects);

(vi) compilation of a 1:1 million scale computerized land resources data base (agro-ecological cells) of each district in a Geographic Information System (GIS);

(vii) determination of land under other uses;

(viii) determination of land available for productivity assessment of crop, livestock and fuelwood;

(ix) formulation of crop productivity model and assessment of land productivity potential for crop production;

(x) formulation of livestock productivity model and assessment of land productivity potential for pasture and livestock production;

(xi) formulation of fuelwood productivity model and assessment of land productivity potential for fuelwood production;

(xii) assessments of reference land productivity potentials (from crop livestock and fuelwood) for each agro-ecological cell;

FIGURE 2.1
Schematic presentation of methodology

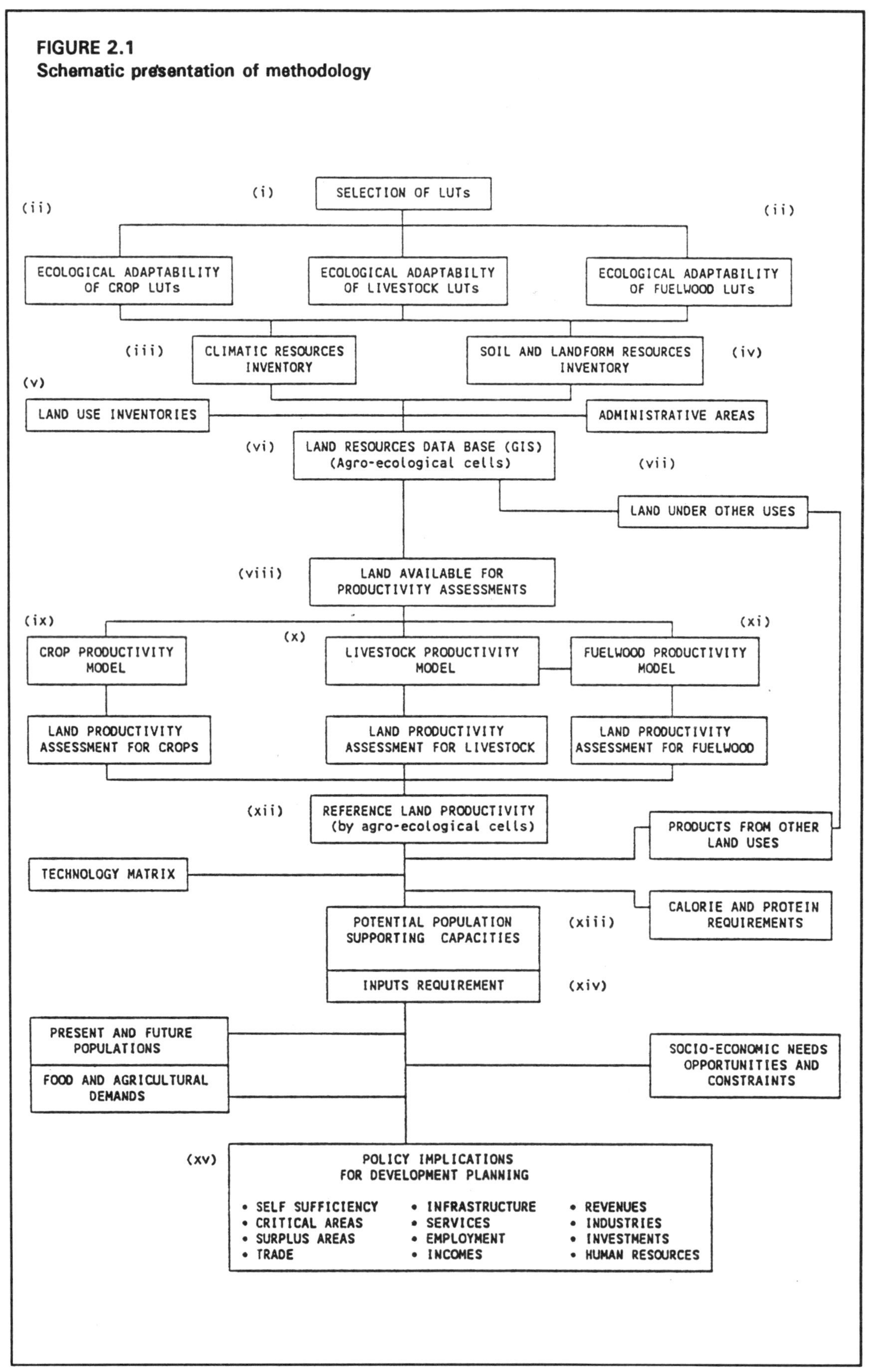

(xiii) assessment of potential population supporting capacity, taking into account nutritional requirements;

(xiv) estimation of inputs requirements, using an input technology matrix including soil conservation; and

(xv) addressing a range of policy issues for development planning, based on a set of scenarios embodying present and future populations, food and agricultural demands, and socio-economic development needs, opportunities and constraints.

The above 15 activities represent four groups of compound activities namely:

(a) formulation of land utilization types and their ecological requirements;

(b) compilation of national land resources and land use data base;

(c) assessments of land productivity potentials; and

(d) development planning involving assessments of potential population supporting capacities and input requirements to address policy issues.

Activities related to the formulation of land utilization types and their ecological requirements overlapped with those activities concerned with the compilation of the land resources data base. This was necessary to ensure that those land qualities that determine productivity characterized in the land inventory and that land use requirements can be formulated in terms of such land qualities.

Subsequently, the productivity models of crops, livestock and fuelwood were applied to the land resources data base to estimate land productivity potentials of alternative kinds of land uses (land utilization types). These land productivity potentials, in turn, form a basis for quantifying potential population supporting capacities and input requirements at several levels of geographical and administrative aggregation (e.g. sub-district, district, province, nation). When set against present and projected population distribution, food and agricultural demands, socio-economic development needs, opportunities and constraints, such assessments of land and population potentials provide a sound and coherent basis for national development planning.

2.2 Land Utilization Types and Ecological Requirements

In the crop productivity model (Chapter 5), a total of 25 crop species are considered. They are differentiated into 64 crop types to account for differences in ecotype adaptation, crop phenology and growth cycles within each species. The crops comprise:

7 cereal food grain crops (wheat, barley, oat, rice, maize, pearl millet, sorghum)
6 leguminous food crops (cowpea, green gram, groundnut, phaseolus bean, pigeonpea, soybean)
3 root and tuber crops (white potato, sweet potato, cassava)
9 'cash' crops (banana, oil palm, sugarcane, coffee, cotton, pineapple, pyrethrum, sisal, tea).

Coffee, cotton, pineapple, pyrethrum, sisal and tea are considered in the model to take account of the reported land area occupied by these crops as inventoried in the land use data base. The remaining 19 crops (58 crop types) are considered at three levels of inputs circumstances. In all, therefore, the crop productivity model considers 180 crop land utilization types (LUT).

In the livestock productivity model (Chapter 6), a total of 30 pasture and fodder species of grasses and legumes, and six livestock types have been considered. The grass and fodder species comprise:

18 pasture grass species
4 fodder grass species
8 pasture and fodder legumes.

The above grass and legume species are considered at three levels of inputs circumstances, and the livestock model has a provision for considering 90 pasture and fodder LUTs. However, at this stage of model development and application, productivity assessments have been made for groups of pasture and fodder grasses and legumes at three levels of inputs.

The six livestock types considered in the livestock productivity model are: cattle, goat, sheep, camel, poultry and pig, of which the first four are considered under non-pastoral as well as pastoral systems. Poultry and pig are considered without explicitly defining the production systems. The non-pastoral and pastoral systems comprise:

Non-pastoral:
3 Dairy and meat cattle systems
3 Dairy and meat goat systems
3 Meat and wool sheep systems.

Pastoral:
3 Cattle herds (nomadic distant, nomadic with market access, semi-nomadic)
2 Goat herds (nomadic distant, semi-nomadic)
2 Sheep herds (nomadic distant, semi-nomadic)
1 Camel herd (nomadic).

The livestock model makes a provision for a total of 27 livestock LUTs so that the above livestock systems, including those of poultry and pig, can be assessed at three levels of inputs circumstances.

In the fuelwood productivity model (Chapter 7), 31 species of fuelwood are considered. They comprise:

15 species with nitrogen fixing ability
16 species without nitrogen fixing ability.

The model considers the 31 fuelwood species at three levels of inputs circumstances, so that the fuelwood model has a provision for assessing 93 fuelwood LUTs.

The three levels of inputs circumstances used in the models are: low inputs, intermediate inputs and high inputs as detailed in Chapters 5, 6 and 7. The low level circumstance assumes low capital input and subsistence management practices, the use of 'indigenous' cultivars of crops and breeds of animals, hand labour only, no use of fertilizer or biocides, no large-scale conservation measures, and cultivation in rotation with bush fallow to maintain soil fertility. It can be compared to traditional systems of bush fallow rotations. The intermediate level circumstance assumes medium capital input, partly subsistence and partly commercial management practices, the use of improved cultivars of crops and breeds of animals (including crossbred animals), use of improved hand tools and draught implements, some mechanization, some use of fertilizer and biocides, some soil conservation measures, and cultivation in rotation with sown grass fallow. The high level circumstance assumes capital intensive management practices, full use of most productive adapted cultivars of crops and breeds of animals (including exotic breeds), complete mechanization, optimum use of farm chemicals, and full soil conservation measures.

Determination of the climatic and edaphic (soil) requirements of crop, livestock and fuelwood land utilization types used in the assessment has been a major activity. Previous attempts to quantify climatic requirements of crops (including pasture and fuelwood) have not adequately recognized the importance of the time courses of temperature and soil moisture balance (including seasonal and between year variations) in relation to crop growth (photosynthesis), development (phenology) and production (yield). Adequate emphasis has been placed on these two sets of parameters (temperature and soil moisture regimes) in this assessment.

Of similar significance is the nature of the photosynthesis response to temperature and radiation, which determines crop yield and land productivity when the phenological requirements are met during the period when soil moisture is available for growth. Accordingly, an inventory of crop, pasture and fuelwood species was prepared, based on their climatic requirements for both photosynthesis and phenology. Four main climatic adaptability groups of crops, pasture grasses and legumes, and fuelwood species are recognized in the assessment, as detailed in Chapters 5, 6 and 7. This inventory gives, among other information, ranges of temperature requirements for different aspects of growth and development. These are subsequently matched to existing thermal climatic conditions.

Once the photosynthetic and phenological thermal requirements are met, the agronomic (or silvicultural) yield potential of a crop, under constraint-free conditions, is governed by the number of days (or years) to maturity. This, in turn, is determined by the length and quality of growing period (including its year-to-year variation). Constraint-free yields were calculated for all crop, livestock and fuelwood LUTs for each length of growing period. The results were used as the basis of the climatic suitability assessment, as described in Chapters 5, 6 and 7.

Soil requirements of LUTs were assessed as follows: For each crop, pasture and fuelwood species, available data on soil characteristics considered meaningful for production were listed, e.g. soil depth, texture, salinity, stoniness, etc. For each LUT, each property was then quantitatively subdivided into those for optimum conditions and for acceptable range of conditions. When a property fell outside the defined range, the soil was considered as currently not suitable.

The information on optimal range and minimal or maximum values of soil properties for each LUT formed the basis for subsequent suitability rating of the soil units for production of crops, pasture and fuelwood (Chapters 5, 6 and 7).

2.3 Compilation of National Land Resources and Land Use Database

The information on climatic and soil requirements of crop, pasture and fuelwood land utilization types, was used as a guide in the compilation of the land resources inventory (Chapter 3).

In the case of climate, temperature and soil moisture availability are key factors in determining the distribution (in both space and time) of rainfed crops. In combination with solar radiation, these climatic factors condition photosynthesis and allow plants to accumulate biomass (and accomplish successive development stages) according to their ecophysiological rates and patterns.

The temperature (heat) attributes were quantified by defining thermal zones. To cater for the differences in temperature requirements of crops, pasture and fuelwood species, nine reference thermal zones have been inventoried based on 2.5 °C interval in daily mean temperature.

The moisture attributes were quantified through the concept of the reference length of growing period, defined as the duration (in days) of the period during which the supply of available soil moisture from precipitation, and from storage in the soil profile (set at a reference 100 mm), is greater than half the potential evapotranspiration.

Lengths of growing periods were computed from historical data sets of some 435 stations, and derived from average data sets of some 1500 stations. With the historical data set, length of growing periods were computed for individual years, and frequency distributions for each mean length were computed for the historical series. Where there were more than one length of growing period per year, the total mean length as well as the individual mean lengths (e.g. two, three) and their frequency distributions were calculated. These computations represented the information on the length of growing period (LGP). Fifteen mean length of growing period zones, at 30 day interval, have been delineated in the climatic inventory of Kenya, as explained in Chapter 3.

To inventory the year-to-year variation in the number of lengths of growing periods per year, a historical profile was compiled showing groups of years each with a different number of growing periods per year. The proportional representation of each group in the total historical series was computed. This information represents the pattern of length of growing period (LGP-Pattern). Twenty-two LGP-Pattern zones have been recognized in the climatic inventory of Kenya.

For each LGP zone delineated, average values of major climatic elements (radiation, day and night temperature, humidity, etc.) were inventoried to characterize the climate during the growing period. These together with the information on the year-to-year variation in the number of length of growing periods per year and in each component length of growing period, formed the basis for subsequent matching and productivity estimation. Details of the climatic resources inventory are given in Chapter 3.

The soil inventory was compiled essentially from the 1:1 million scale Exploratory Soil Map of Kenya, which comprises 390 different soil map units. For each map unit, information on landform, geology/parent material, soil unit (with implied characteristics), slope-gradient, soil texture and soil phases, in terms of description, classes and extents was transferred to form the soil resources inventory of this assessment (Chapter 3).

On completion of the climatic inventory, the three layers (thermal zone, LGP zone and LGP-Pattern zone) were superimposed on the Exploratory Soil Map of Kenya. The different layers of climate and soil information were digitized and the information was converted to a data base of about 575 thousand one millimeter square grid cells, each corresponding to 100 ha. The resultant map output created about 91 000 unique agro-ecological cells of the inventory, whose land attributes, defined by climate, soil and landform, are known and quantified. This information, compiled at the national level by province and district, constitutes the physical land resources data base of Kenya.

Additional six layers of information were also digitized and overlaid on the land resources inventory. These layers contain information on cash crop zones, forest zones, parkland areas, irrigation schemes, tse-tse infestation areas and administrative boundaries (provinces and districts).

The climate, soil and land use inventories make-up the computerized data base for the assessment, and allow any desired geographical and administrative aggregation to be made of inventoried parameters and results.

2.4 Assessment of Potential Land Productivity

The assessment of land productivity starts by formulating and selecting crop, livestock and fuelwood land utilization types (shown at the head of the flow chart in Figure 2.1), and their ecological (climate, soil and landform) requirements (ii).

Then, from the agro-ecological cells in the land resources inventory (iii, iv, v, vi), district by district, land used or required for irrigation, cash crops and for non-agricultural purposes (vii) is deducted. The remainder is an inventory of land potentially available for rainfed cultivation, and for productivity assessments (viii).

For each of the agro-ecological cells in this inventory, the next stage is to determine the potential rainfed yield or output of crops, livestock and fuelwood at one or more levels of inputs (ix, x, xi) in order to find out which land utilization types (cropping patterns and rotations, livestock systems, fuelwood land uses) are most productive, stable and sustainable in the unique conditions of the cell. The land productivity potentials can then be calculated (xii), either in a reference manner or within the context of a set of planning scenarios.

The crop, livestock and fuelwood models (Chapters 5, 6 and 7) are all designed to operate on the computerized land resources data base. They permit quantitative land suitability assessments to be made based on growth and yield predictions of each LUT and combinations of LUTs in each agro-ecological cell. All three productivity models include a provision for quantifying soil erosion hazard of each LUT in terms of productivity loss. This is achieved through the soil erosion and productivity model described in Chapter 4. The model also estimates 'tolerable' soil loss, and costs of alternative conservation measures.

The crop productivity model (Chapter 5) explicitly formulates options in respect of individual crops, annual cropping patterns and crop rotations, and quantifies their production potentials. The model formulates optimum cropping patterns and output therefrom to meet a reference or given food demand, taking into account desired level of production stability.

The livestock productivity model (Chapter 6) quantifies primary productivity potential which is then converted into secondary production (milk, meat, wool, draught power) for pastoral and non-pastoral herds.

The fuelwood model (Chapter 7) quantifies wood biomass productivity potential in terms of mean annual increments over the rotation age of each fuelwood LUT.

The crop, livestock and fuelwood productivity models are interphased with each other. This allows land productivity to be optimized for a given set of development constraints and demands.

2.5 Development Planning

Beyond this, the assessment allows for development planning applications, which involve the calculation of the quantities of edible calories and protein that would be produced by the different crops and livestock (and products from other land uses) from information on the nutritional composition of the products. The crops or crop mixes (including grassland) that can produce the largest or desired quantity and quality of calories and protein in each agro-ecological cell are then selected, and the results from each cell in each climatic zone in each district are added to determine the optimal maximum potential production of calories and protein from each climatic zone in each district, from whole districts and groups of districts, and from whole provinces and country.

Dietary and other constraints such as minimum protein requirements are applied to estimate potential population supporting capacity (xiii) at various desired levels of geographical and administrative aggregation. Similarly, by applying the extended FAO technology matrix for Kenya (including conservation inputs), the associated inputs requirements (xiv) are quantified (Bruinsma, Hrabovszky, Alexandratos and Petri, 1983).

The potential population supporting capacity (in xiii) is computed as potential population density (persons per ha) which is compared with the present and projected population densities, and examined against food and agriculture demands, and socio-economic needs, opportunities and constraints, to address a range of policy issues for development planning. These relate, for example, to food and economic self-sufficiency, areas with surplus potential and areas that are critical, domestic and export trade, infrastructure, services, employment, incomes, revenues, industries, investments and human resources development (xv).

Chapter 3

Land resources

The land resources inventory brings together two layers of information on physical environmental resources (climate and soil) and allows the creation of unique ecological land units (agro-ecological cells) within which soil, landform and climatic conditions are quantified. This information, compiled at the national level by province and district, constitutes the inventory of the physical land resources.

To create a computerized inventory of land resources, the individual climate and soil inventories have been compiled in map form at 1:1 million scale, and digitized.

The climatic resources inventory consists of three seperate thematic layers: thermal zones, length of growing period zones, and pattern of number of length of growing period zones. The Exploratory Soil Map of Kenya (KSS 1982a), forms the soil base, providing information on soils, landform and geology/parent material.

Six additional layers (1:1 million scale) of information have also been digitized and overlaid on the land resources inventory. These layers provide information on cash crop zones, forest zones, parkland areas, irrigation schemes, tse-tse infestation areas, and province and district boundaries.

The individual map layers have been digitized using the Comprehensive Resource Inventory and Evaluation System (CRIES 1983) a GIS developed at Michigan State University. The digitized information derived from the individual map layers has been converted to a data base of 576,072 grid cells. Each cell (one millimeter square) corresponds to 100 ha.

Subsequent to digitizing the individual layers of the land resources inventory, the soil mapping unit composition of each mapping unit and the associated ecological conditions have been incorporated.

The make-up of the national land resources data base is schematically presented in Figure 3.1, and is described in Technical Annex 1. A map of provinces and districts is presented in Figure 3.2, and their extents are given in Table 3.1.

3.1 Climate Resources

Temperature and water are the major climatic factors that govern crop distribution (both in space and time). In combination with solar radiation, these climatic factors condition the net

FIGURE 3.1
Make-up of land resources data base

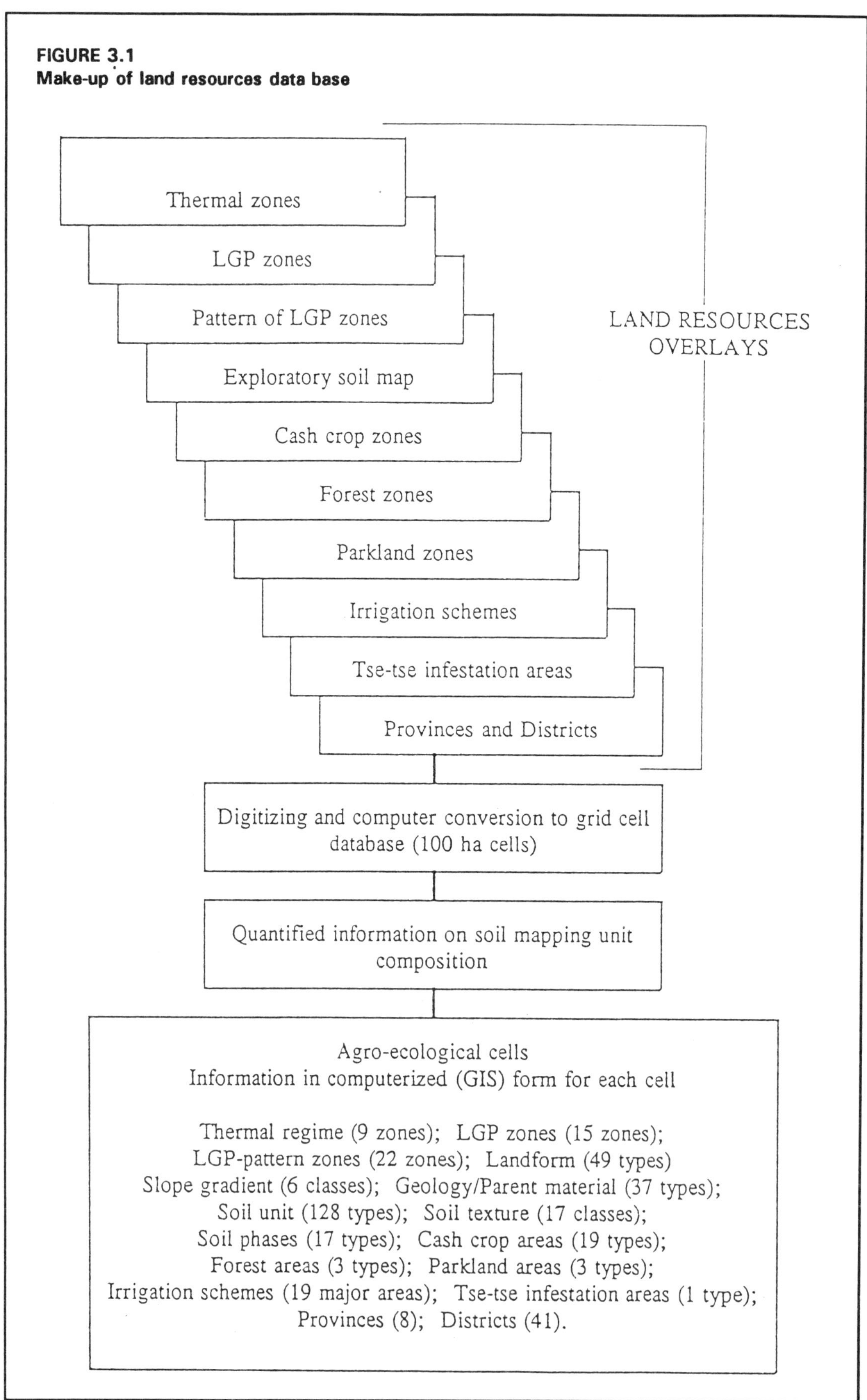
Thermal zones
LGP zones
Pattern of LGP zones
Exploratory soil map
Cash crop zones
Forest zones
Parkland zones
Irrigation schemes
Tse-tse infestation areas
Provinces and Districts
LAND RESOURCES OVERLAYS
Digitizing and computer conversion to grid cell database (100 ha cells)
Quantified information on soil mapping unit composition
Agro-ecological cells
Information in computerized (GIS) form for each cell
Thermal regime (9 zones); LGP zones (15 zones);
LGP-pattern zones (22 zones); Landform (49 types)
Slope gradient (6 classes); Geology/Parent material (37 types);
Soil unit (128 types); Soil texture (17 classes);
Soil phases (17 types); Cash crop areas (19 types);
Forest areas (3 types); Parkland areas (3 types);
Irrigation schemes (19 major areas); Tse-tse infestation areas (1 type);
Provinces (8); Districts (41).

FIGURE 3.2
District map of Kenya

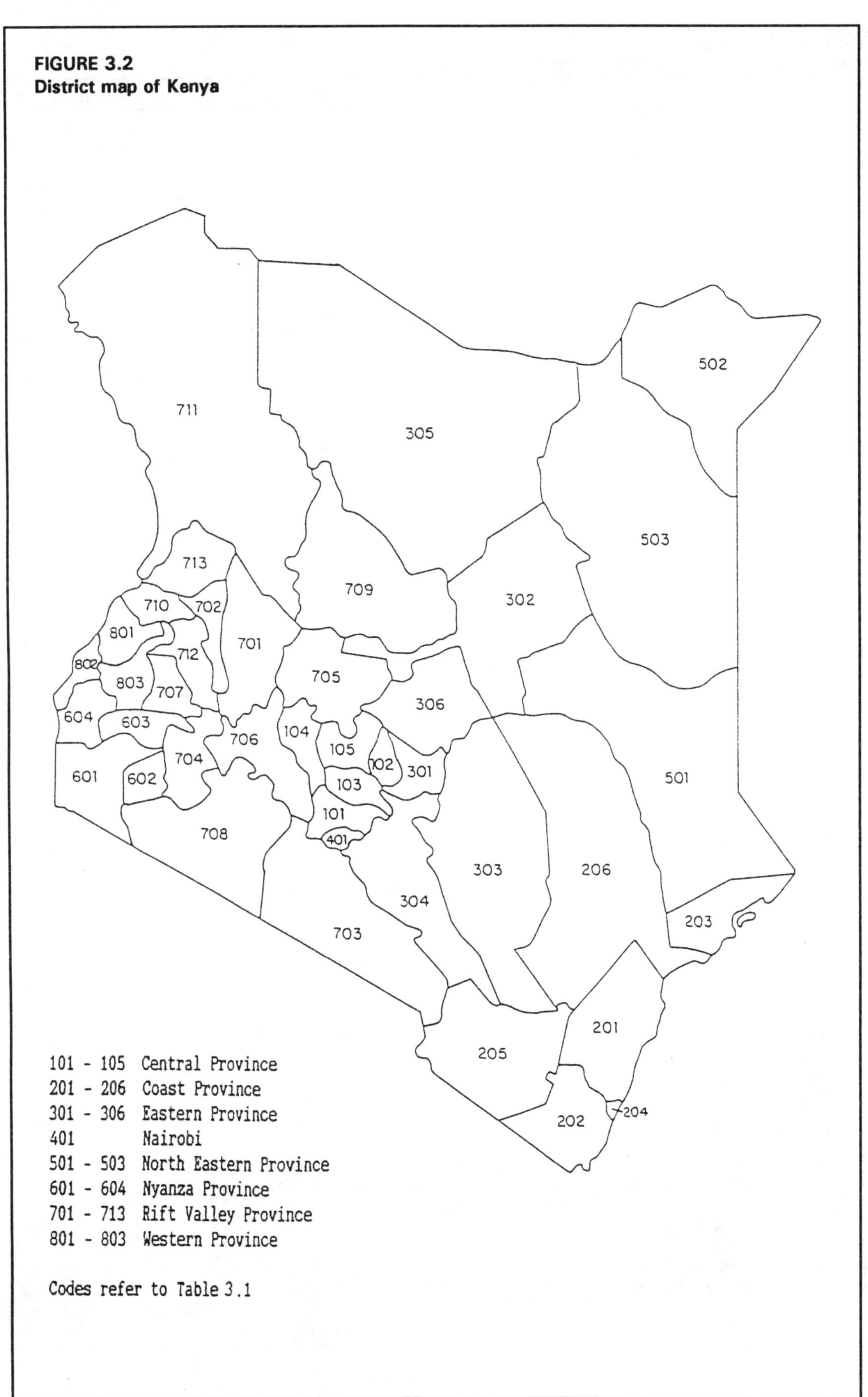
711
305
502
503
709
713
302
710
702
801
701
712
802
803
705
707
306
604
603
706
104
105
102
301
704
601
602
103
501
101
708
401
303
206
304
203
703
201
205
202
204
101 - 105 Central Province
201 - 206 Coast Province
301 - 306 Eastern Province
401 Nairobi
501 - 503 North Eastern Province
601 - 604 Nyanza Province
701 - 713 Rift Valley Province
801 - 803 Western Province
Codes refer to Table 3.1

TABLE 3.1
Extents of Districts and Provinces of Kenya

Code	Name	'000s ha	percent of total
101	Kiambu	256.7	0.45
102	Kirinyaga	138.6	0.24
103	Muranga	254.6	0.44
104	Nyandarua	333.4	0.58
105	Nyeri	339.0	0.59
	Central Province	**1 322.3**	**2.30**
201	Kilifi	1 262.5	2.19
202	Kwale	814.4	1.41
203	Lamu	654.0	1.14
204	Mombasa	25.4	0.04
205	Taita	1 735.2	3.01
206	Tana River	3 939.2	6.84
	Coast Province	**8 430.9**	**14.63**
301	Embu	271.2	0.45
302	Isiolo	2 542.1	4.41
303	Kitui	3 037.2	5.27
304	Machakos	1 463.1	2.54
305	Marsabit	7 087.0	12.31
306	Meru	976.7	1.70
	Eastern Province	**15 377.3**	**26.69**
401	Nairobi	75.4	0.13
	Nairobi	**75.4**	**0.13**
501	Garissa	4 408.7	7.65
502	Mandera	2 617.8	4.55
503	Wajir	5 722.9	9.93
	North Eastern Province	**12 749.4**	**22.13**
601	South Nyanza	588.5	1.02
602	Kisii	214.2	0.37
603	Kisumu	217.4	0.38
604	Siaya	247.2	0.43
	Nyanza Province	**1 267.3**	**2.20**
701	Baringo	1 066.0	1.81
702	Elgeyo Mar.	263.0	0.46
703	Kaijado	2 151.9	3.74
704	Kericho	478.1	0.83
705	Laikipia	927.5	1.61
706	Nakuru	742.7	1.29
707	Nandi	281.4	0.49
708	Narok	1 799.1	3.12
709	Samburu	2 051.1	3.57
710	Trans Nzoia	246.7	0.43
711	Turkana	6 586.8	11.43
712	Uasin Gishu	382.6	0.66
713	West Pokot	535.7	0.93
	Rift Valley Province	**17 512.6**	**30.40**
801	Bungoma	319.3	0.56
802	Busia	191.4	0.33
803	Kakamega	361.5	0.63
	Western Province	**872.2**	**1.52**
	KENYA	**57 607.2**	**100.00**

photosynthesis and allow plants to accumulate dry matter (and to accomplish the successive development stages), according to the rates and patterns which are specific to cultivated plants.

The growing period has been used as a framework for the assessment of climatic resources (FAO 1978-81). It is defined as the period in which temperature and moisture permit crop growth. Prevailing temperature regimes have been inventoried by identification of thermal zones in order to take into account temperature requirements of crops (including pasture and fuelwood species).

The inventory of climatic resources allows:

(i) A differentiation of the country into reference thermal zones, reflecting the geographical and 'seasonal' distribution of the prevailing temperature regimes.

(ii) A differentiation of the country into reference length and pattern of growing period zones, reflecting the prevailing moisture regimes including the year-to-year variations.

(iii) A quantification of potential yields (of crops, livestock and fuelwood) that can be attained under constraint-free conditions.

(iv) An assessment of various agro-climatic constraints to take into account yield losses likely to occur.

The climatic data bank, growing period analysis and the creation of the climatic resources inventory are described in Technical Annex 1.

3.1.1 Climatic Data Bank

The climatic data[1] bank compiled for the assessment consists of three data sets. Data set 1 (Historical Data, from Jaetzhold and Kutsch 1980) consists of the following information for 437 stations:

- Average decadal (10-day total) potential evapotranspiration (mm)
- Historical decadal rainfall (mm) for individual years.

Data set 2 (Average climatic data, from FAO/AGPC data bank, FAO 1977) consists of 45 stations with mean monthly values for the following 11 climatic parameters:

- Precipitation (mm)
- Mean daily temperature (°C)
- Maximum temperature (°C)
- Minimum temperature (°C)
- Day-time temperature (°C)
- Night-time temperature (°C)
- Mean water vapour pressure (mbar)
- Mean wind velocity (m sec^{-1})
- Hours of bright sunshine as a percentage of maximum possible sunshine hours (%)

---

1 The primary source of the climatic data is the Kenya Meteorological Department.

- Solar radiation (cal cm^{-2} day^{-1})
- Potential evapotranspiration (mm).

Data set 3 (Average climatic data, from Kenya Soil Survey) provides average data on the following for 1489 stations:

- Annual daily temperature (°C)
- Annual potential evaporation (mm)
- Annual potential evapotranspiration (mm)
- Annual Rainfall (mm)
- Monthly rainfall (mm)
- Type of rainfall pattern; monomodal (M), bimodal (B) or trimodal (T).

Extracts of the three data sets are presented in Tables 3.2, 3.3 and 3.4 respectively. The complete climatic data bank is available on diskettes (ASCII).

3.1.2 Growing Period Model

The definitions and model used to quantify the reference length of growing period have been described in Technical Annex 1. The growing period is the time period when moisture supply exceeds half potential evapotranspiration; it includes the time required to evapotranspire up to 100 mm of soil moisture storage[2]. The calculation of the reference growing period is based on a water balance model, comparing rainfall with potential evapotranspiration. The length of growing period (and the number of growing periods and dry periods per year) from a climatic viewpoint alone, and independent of crop, soil and landform, is therefore quantified in a reference manner (Kowal and Kassam 1978; Doorenbos and Kassam 1979; Kassam, van Velthuizen, Higgins, Christoforides, Voortman and Spiers 1982; Brammer, Antoine, Kassam and van Velthuizen 1988).

Two types of growing periods are schematically shown in Figure 3.3. The distinction between 'normal' and 'intermediate' is useful because in the latter it is unlikely that full water requirements can be met during the rainy season without moisture conservation or a supply from groundwater or irrigation.

Two more growing period types have been identified (Figure 3.3). These are: (a) all year-round humid with rainfall exceeding full potential evapotranspiration throughout the year, and (b) all year-round dry with rainfall not exceeding half potential evapotranspiration throughout the year.

3.1.2.1 Length of Growing Period (LGP)

Mean length of growing period and frequency distribution for each individual group of years have been computed. Where there are more than one length of growing period per year, the total mean length as well as the individual mean lengths (e.g. two, three) and their frequency distribution are calculated (Figure 3.4).

---

2 The computer program is able to handle a storage term in the range 0-250 mm.

TABLE 3.2
Extract agroclimatic data bank - Data set 1 - Historical data

STATION: EMBU NUMBER: 9037008 LAT: 0.32° S LONG: 32.27° E ALT: 1410 FT 40 YEARS' RECORDS

		Jan	Feb	Mar	Apr	May	Jun	Jul	Aug	Sep	Oct	Nov	Dec	Year
AVERAGE PET	DECAD 1	40.8	43.2	44.8	40.0	37.6	34.4	32.8	32.8	37.6	40.8	36.8	36.8	
	DECAD 2	42.4	43.2	45.6	37.6	37.6	33.6	32.0	32.8	39.2	41.6	35.2	36.8	
	DECAD 3	42.4	44.0	44.0	37.6	36.8	33.6	32.0	34.4	40.0	40.0	36.0	38.4	
	MONTH	125.4	130.4	134.4	115.2	112.0	101.6	96.8	100.0	116.8	122.4	108.0	112.0	1375.0
RAINFALL (1927)	DECAD 1	1.8	0.0	29.0	20.0	47.3	0.3	8.1	2.9	6.6	0.0	19.3	14.5	
	DECAD 2	0.0	2.0	167.8	64.5	19.1	52.2	2.1	7.0	0.0	58.3	68.1	16.8	
	DECAD 3	0.0	1.5	52.3	75.3	3.1	0.3	0.8	2.8	4.1	82.8	8.9	0.0	
	MONTH	1.8	3.5	249.1	159.8	69.5	52.8	11.0	12.7	10.7	141.1	96.3	31.3	839.6
RAINFALL (1928)	DECAD 1	15.6	0.0	40.9	10.2	101.1	16.2	0.0	2.6	0.0	6.4	40.7	21.9	
	DECAD 2	39.1	0.0	0.0	140.2	103.0	1.1	6.7	0.0	0.0	0.0	108.2	0.0	
	DECAD 3	0.0	0.0	7.1	80.4	62.1	3.8	0.0	5.6	5.1	17.3	49.2	0.0	
	MONTH	54.7	0.0	48.0	230.8	266.2	21.1	6.7	8.2	5.1	23.7	198.1	21.9	884.5
RAINFALL (1929)	DECAD 1	51.6	0.0	0.0	12.4	66.7	0.8	10.9	0.0	32.8	1.0	108.7	102.9	
	DECAD 2	6.6	0.0	66.5	12.4	7.9	7.9	14.3	0.0	27.2	34.0	104.5	38.0	
	DECAD 3	6.4	0.0	5.1	153.5	14.3	2.0	12.5	5.8	0.0	80.8	4.6	12.3	
	MONTH	64.6	0.0	71.6	178.3	88.9	10.7	37.7	5.8	60.0	115.8	217.8	153.2	1004.4
RAINFALL (1930)	DECAD 1	0.0	28.7	76.5	57.8	66.9	12.5	0.0	10.7	2.6	26.7	122.8	42.9	
	DECAD 2	0.0	0.0	20.6	109.7	28.3	0.0	7.6	6.1	5.6	28.4	94.7	28.0	
	DECAD 3	43.0	0.0	80.7	107.3	5.4	6.6	0.8	33.0	2.5	66.1	53.9	1.0	
	MONTH	43.0	28.7	177.8	274.8	100.6	19.1	8.4	49.8	10.7	121.2	271.4	71.9	1177.4

TABLE 3.3
Extract agroclimatic data bank - Data set 2 - Average climatic data

COUNTRY: KENYA STATION: LOKITAUNG NUMBER: 63610 LAT: 4.15° LONG: 35.45°E ELEVATION: 730 m

	Jan	Feb	Mar	Apr	May	Jun	Jul	Aug	Sep	Oct	Nov	Dec	Year
PRECIPITATION (mm)	12	17	53	119	47	22	32	13	7	13	32	28	395
TEMPERATURE (°C; AVERAGE)	28.0	28.5	28.2	26.0	26.6	26.6	25.7	25.7	26.8	26.8	27.1	26.8	26.9
TEMPERATURE (°C; MEAN MAX.)	33.2	33.8	33.2	31.0	31.0	31.0	30.5	30.5	31.6	31.6	32.1	31.6	31.8
TEMPERATURE (°C; MEAN MIN.)	22.7	23.2	23.2	21.0	22.1	22.1	21.0	21.0	22.1	22.1	22.1	22.1	22.1
TEMPERATURE (°C; MEAN DAY)	29.8	30.4	30.0	27.8	28.1	28.1	27.5	27.5	28.5	28.5	28.9	28.5	28.6
TEMPERATURE (°C; MEAN NIGHT)	26.1	26.6	26.4	24.2	24.9	24.9	24.0	24.0	25.1	25.1	25.2	25.1	25.1
VAPOUR PRESSURE	21.1	22.7	23.0	23.0	22.5	20.8	20.5	20.1	20.1	21.6	23.0	22.0	21.7
WIND SPEED (at 2m ELEV.)	3.7	4.3	4.0	2.9	3.3	3.7	3.8	3.7	4.1	4.1	4.3	4.0	3.8
SUNSHINE (%)	84	81	75	78	83	83	76	83	88	84	76	81	81
TOTAL RADIATION	530	548	545	552	550	535	515	557	587	563	510	510	541
EVAPOTRANSPIRATION	188	184	198	157	171	171	169	178	194	191	174	173	2148

COUNTRY: KENYA STATION: LODWAR NUMBER: 63612 LAT: 3.07° LONG: 35.37°E ELEVATION: 515 m

	Jan	Feb	Mar	Apr	May	Jun	Jul	Aug	Sep	Oct	Nov	Dec	Year
PRECIPITATION (mm)	15	8	27	56	27	6	23	10	2	9	21	16	220
TEMPERATURE (°C; AVERAGE)	28.8	29.8	30.2	29.6	29.7	29.0	28.2	28.5	29.3	29.8	29.0	28.6	29.2
TEMPERATURE (°C; MEAN MAX.)	35.5	36.5	36.0	35.0	34.8	34.1	33.0	33.3	34.8	35.2	34.6	34.5	34.8
TEMPERATURE (°C; MEAN MIN.)	22.2	23.1	24.3	24.3	24.6	24.0	23.5	23.5	23.8	24.5	23.5	22.6	23.7
TEMPERATURE (°C; MEAN DAY)	31.2	32.2	32.3	31.6	31.5	30.9	30.0	30.2	31.3	31.8	31.0	30.7	31.2
TEMPERATURE (°C; MEAN NIGHT)	26.5	27.4	28.1	27.7	27.8	27.2	26.5	26.6	27.3	27.9	27.0	26.3	27.2
VAPOUR PRESSURE	17.0	17.0	19.3	22.0	22.0	20.3	19.5	19.5	18.7	18.7	19.0	19.0	19.3
WIND SPEED (at 2m ELEV.)	2.6	2.8	3.0	2.8	2.8	2.6	2.4	2.8	3.0	3.2	2.8	2.4	2.8
SUNSHINE (%)	84	81	75	78	83	83	76	83	88	84	76	81	81
TOTAL RADIATION	537	553	546	550	545	529	509	553	587	567	516	517	542
EVAPOTRANSPIRATION	197	195	205	186	188	171	167	186	198	208	178	170	2249

TABLE 3.4
Extract agroclimatic data bank - data set 3 - Average data

Station code	Station name	°Lat (N/S)	°Long (E)	Alt (ft)	Average Annual Data				Average Monthly Rainfall Data												Yrs	RP type
					Temp	Eo	PET	P	Jan	Feb	Mar	Apr	May	Jun	Jul	Aug	Sep	Oct	Nov	Dec		
9439000	Kilindini	4.03	39.39	64	26	2168	1734	1059	28	13	56	171	262	106	68	64	66	88	83	54	50	B
9439001	Kwale Agr. Dept.	4.11	39.28	1294	24	2075	1662	1089	34	19	60	159	227	98	81	64	67	99	102	79	60	B
9439002	Mombasa Old.	4.04	39.41	53	26	2169	1735	1193	26	16	62	199	313	113	86	67	68	87	99	57	81	B
9439003	Ramisi Ass. Sug.	4.31	39.25	50	26	2175	1740	1426	23	18	78	271	359	155	121	82	61	84	103	71	39	B
9439004	Gazi Kenya Sug.	4.25	39.30	150	26	2161	1729	1350	23	24	70	256	347	150	100	86	68	79	89	58	50	B
9439005	Waa	4.10	39.37	68	26	2175	1740	1289	28	10	71	240	340	118	28	77	122	90	96	69	10	B
9439008	Mrere Works	4.12	39.24	650	26	2161	1729	1049	26	20	52	109	255	102	71	70	69	104	98	73	23	B
9439009	Changamwe	4.02	39.38	200	26	2153	1722	1093	8	15	47	182	308	94	60	71	70	96	71	71	23	B
9439010	Msumbweni Hosp.	4.30	39.30	62	26	2171	1573	1376	20	16	62	293	356	142	111	79	66	93	86	52	30	B
9439013	Vanga Mudir's	4.40	39.13	40	26	2171	1737	1134	26	19	86	198	274	88	79	67	51	86	98	62	36	B
9439014	Gazi Mudir's	4.28	39.29	20	26	2175	1740	1388	21	18	69	275	377	133	108	81	66	86	104	50	35	B
9439015	Kwango Mudir's	4.08	39.19	650	26	2193	1754	814	20	33	52	119	150	55	58	49	39	84	105	50	27	B
9439016	Tiwi Disp.	4.14	39.35	30	26	2179	1743	1290	25	17	58	259	327	98	95	82	86	81	93	69	25	B
9439019	Mombasa Met. St.	4.03	39.39	52	26	2169	1735	1202	26	17	63	196	320	120	89	64	64	87	96	60	54	B
9439020	Ros Serani	4.05	39.41	30	26	2172	1737	1241	10	13	42	213	358	92	84	92	80	99	97	61	11	B
9439021	Mombasa Air.	4.02	39.37	185	26	2155	1724	1054	34	19	63	167	231	70	65	67	78	94	97	69	26	B
9439023	Bamburi	4.00	39.43	15	26	2173	1738	1146	0	7	15	228	253	89	150	58	81	118	46	101	3	B
9439024	Mkomani	4.03	39.41	50	26	2170	1736	969	2	1	0	143	178	83	117	71	61	91	91	131	2	B
9439025	Kinango Pump.	4.09	39.25	400	26	2193	1738	930	34	21	55	151	171	67	64	52	48	81	101	85	24	B
9439026	Kisauni	4.02	39.40	50	26	2178	1741	1096	20	8	40	236	275	88	98	62	63	65	80	61	11	B
9439027	Mwangulu	4.25	39.07	400	27	2243	1794	839	24	27	74	112	135	48	54	59	47	70	122	67	12	B
9439028	Ndavaja	4.15	39.10	500	27	2239	1791	810	32	21	61	109	134	47	67	64	42	76	103	54	22	B
9439029	Makuja	4.10	39.34	400	26	2151	1721	947	29	13	47	154	251	57	74	51	47	75	98	51	11	B
9439030	Muhaka	4.20	39.31	150	26	2173	1738	1129	23	19	50	204	260	104	93	79	59	93	105	40	19	B
9439031	Mrore I	4.13	39.25	1335	24	2081	1665	987	34	19	48	150	195	76	63	73	52	87	117	73	22	B
9439032	Mrore II	4.15	39.23	1160	25	2108	1686	1040	36	26	47	161	219	83	83	73	52	99	73	88	12	B
9439033	Mrore III	4.17	39.26	720	25	2139	1711	1094	13	26	53	154	237	93	96	83	50	130	75	84	10	B
9439034	Mrore IV	4.17	39.21	980	25	2132	1706	1077	30	26	45	162	219	84	96	79	50	101	77	108	10	B
9439038	Waa Disp.	4.10	39.35	100	26	2172	1738	1070	24	15	42	191	266	107	79	62	60	80	99	45	18	B
9439040	Mombasa Rest.	4.03	39.39	70	26	2167	1734	1246	54	12	71	267	241	110	86	67	85	81	105	67	10	B
9439041	Mombasa Fields	4.03	39.40	55	26	2169	1735	1145	47	16	58	239	233	114	85	61	75	73	87	57	11	B
9439043	Simba Hills	4.22	39.25	800	25	2120	1696	1290	44	13	71	223	270	106	104	81	66	127	122	63	19	B
9439044	Kikoneni	4.28	39.17	500	26	2172	1738	1242	25	14	92	209	234	117	107	89	70	84	124	77	16	B
9439045	Kimansi W.	4.03	39.39	50	26	2170	1736	1257	28	16	62	192	369	114	87	68	72	88	101	60	78	B
9439046	Vangalunga	4.33	39.07	200	27	2205	1764	911	31	21	77	146	176	61	51	44	38	72	123	71	20	B
9439050	Changamwe	4.01	39.37	100	26	2164	1731	860	31	47	46	87	211	65	80	0	42	83	109	59	2	B
9439051	Timbwani	4.07	39.40	50	26	2175	1740	814	1	0	63	154	228	130	69	41	38	25	24	41	2	B
9439054	Mwena School	4.29	39.08	250	27	2247	1798	752	21	11	75	62	99	90	72	44	38	99	83	58	3	B
9439057	Puma Camp	4.06	39.14	600	27	2226	1781	442	0	0	12	64	127	64	64	22	30	2	2	55	2	B

Stations 8535201 to 8940003 latitude North and stations 9034001 to 9439057 latitude South.
Temp: Mean daily temperature (celcius) - Eo: Potential evaporation (mm) - PET: potential evapotranspiration (mm)
P: Rainfall (mm) - RP type: Rainfall pattern type - M = Monomodal, B = Bimodal, T = Trimodal

FIGURE 3.3
Schematic presentation of types of growing periods

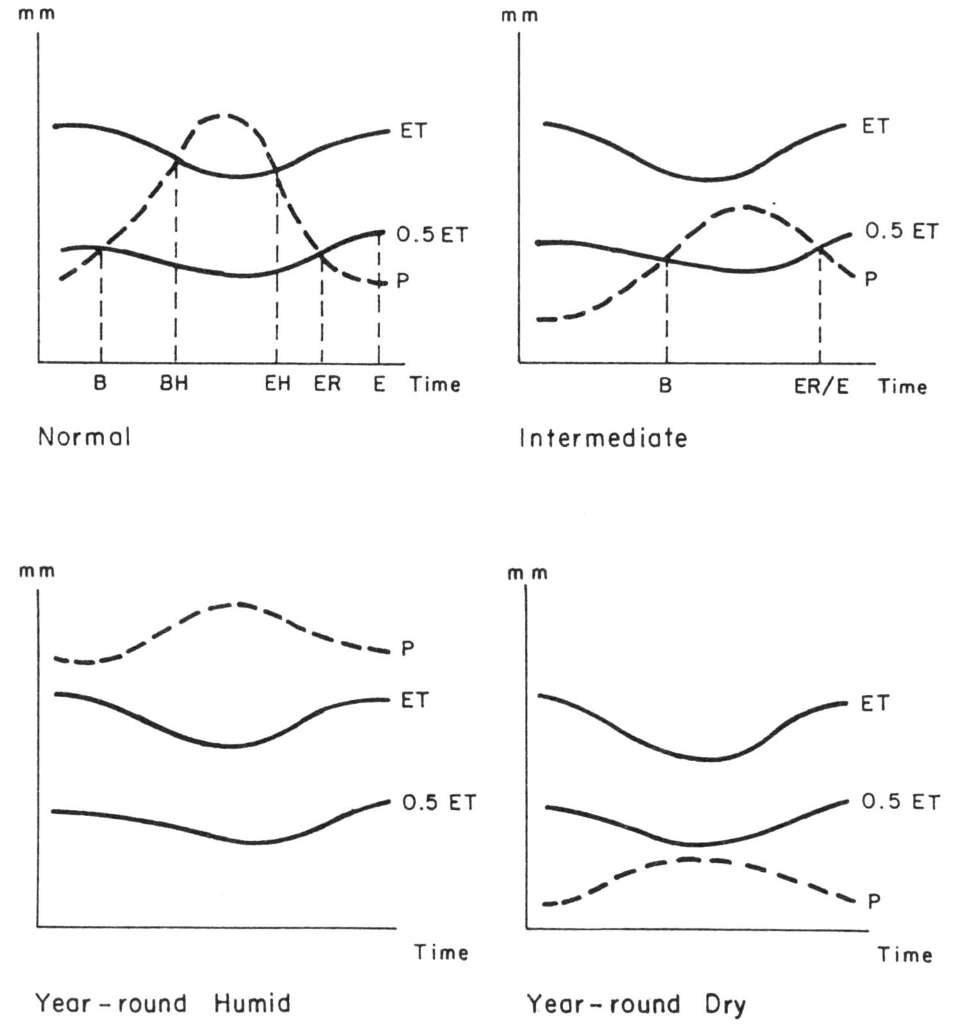

B - Beginning of growing period
BH - Beginning of humid period
EH - End of humid period
ER - End of rainy season
E - End of growing period

P - Precipitation
ET - Potential evapotranspiration

FIGURE 3.4
Number of growing periods and dry periods per year

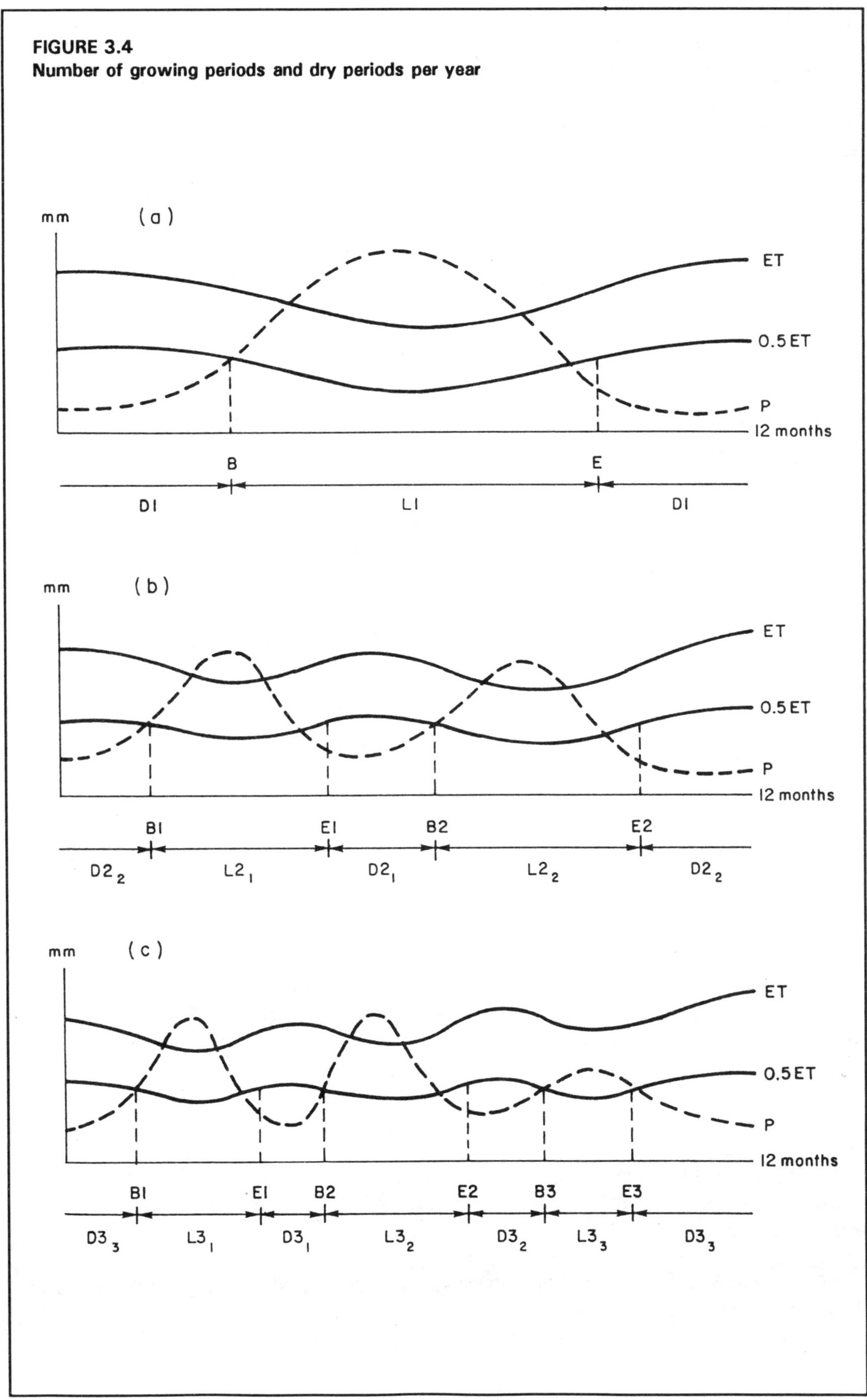

TABLE 3.5
Patterns of growing periods (LGP-patterns) - Historical profiles of occurrence of number of length of growing periods per year

Code	LGP-pattern	Proportion (%)
1	1	100
2	H - 1	60 : 40
3	1 - H	70 : 30
4	1 - H - 2	65 : 20 : 15
5	1 - 2 - H	65 : 20 : 15
6	1 - 2	65 : 35
7	1 - 2 - 3	50 : 35 : 15
8	1 - 3 - 2	40 : 35 : 20
9	1 - 2 - D	40 : 35 : 25
10	1 - D - 2	40 : 35 : 25
11	1 - D	60 : 40
12	2	100
13	2 - 1	70 : 30
14	2 - 1 - H	55 : 30 : 15
15	2 - 1 - 3	55 : 25 : 20
16	2 - 3	75 : 25
17	2 - 3 - 1	60 : 25 : 15
18	2 - 3 - 4	50 : 30 : 10
19	2 - 1 - D	70 : 15 : 15
20	3 - 2	60 : 40
21	3 - 2 - 1	50 : 35 : 15
22	D	100

H = 365^+ days (i.e. year-round humid)
D = zero days (i.e. year-round dry)

TABLE 3.6
Relationships between mean total dominant and mean total associated lengths of growing period

LGP-Pattern	Relationship
1 - 2 1 - 2 - H 1 - H - 2	L2 = 80.40 + 0.75 L1
1 - 2 - 3 1 - 3 - 2 1 - 2 - D 1 - 2 - D	L2 = 71.56 + 0.66 L1 L3 = 77.14 + 0.66 L1
2 - 1 2 - 1 - H 2 - 1 - 3 2 - 1 - D	L1 = -86.09 + 1.28 L2 L3 = 25.29 + 0.82 L2
2 - 3 2 - 3 - 1 2 - 3 - 4	L3 = 30.11 + 0.83 L2 L1 = -98.72 + 1.35 L2 L4 = 114.54 + 0.58 L2
3 - 2 3 - 2 - 1	L2 = 45.05 + 0.80 L3 L1 = -9.86 + 0.88 L3

L1 = Total length of one growing period per year
L2 = Total length of two growing periods per year
L3 = Total length of three growing periods per year
L4 = Total length of four growing periods per year

For a group of years with one length of growing period, the length is designated the code L1, and the dry period is coded D1 (Figure 3.4a). For a group of years with two lengths of growing periods per year, the lengths are coded $L2_1$ and $L2_2$, and the first length ($L2_1$) is followed by the first dry period ($D2_2$) and the second length ($L2_2$) by the second dry period ($D2_2$) (Figure 3.4b) The sum of lengths $L2_1$ and $L2_2$ is coded L2. For a group of years with three lengths of growing periods per year, the lengths are coded $L3_1$, $L3_2$ and $L3_3$, and there are dry periods in between ($D3_1$, $D3_2$, $D3_3$) (Figure 3.4c). The sum of lengths $L3_1$, $L3_2$, $L3_3$ is coded L3.

3.1.2.2 Pattern of Length of Growing Period (LGP-Pattern)

To inventory the year-to-year variation in the number of lengths of growing periods per year, a historical profile is compiled showing groups of years each with a different number of growing periods per year. The proportional representation of each group in the total historical series is computed.

This information represents the pattern of growing period. Twenty two patterns are recognized in the climatic resources inventory. The patterns of number of length of growing period and their composition are presented in Table 3.5.

The pattern of growing period code represents the number of growing periods per year in order of frequency of occurrence, e.g. in the pattern coded as 2-1-3, the numeral 2 represents the number of lengths of growing periods per year (i.e. two) that occur in the majority of the years (i.e. 55 percent) - the dominant length number; the numeral 1 represents number of lengths of growing periods per year (i.e. one) that has the next most commonly occurring frequency (i.e. 25 percent) - the first associated length number; and the numeral 3 represents number of lengths of growing periods per year (i.e. three) that has the smallest occurrence (i.e. 20 percent) - the second associated length number.

For each pattern of growing periods, the mean total length of the dominant number is correlated with the mean total length of the associated numbers. Also, when the mean total length is a summation of more than one mean length, the latter are again correlated with the former. These relationships are presented in Tables 3.6 and 3.7.

TABLE 3.7
Relationship between individual component mean length and mean total length of growing period

LGP-Pattern	Relationship
2 1 - 2 1 - 2 - H 1 - H - 2	$L2_1$ = -1.11 + 0.55 L2 $L2_1$ = 4.94 + 0.62 L2
1 - 2 - 3 1 - 3 - 2 1 - 2 - D 1 - D - 2	$L2_1$ = 5.87 + 0.64 L2 $L3_1$ = 22.12 + 0.39 L3 $L3_2$ = 1.58 + 0.32 L3
2 - 1 2 - 1 - H 2 - 1 - 3 2 - 1 - D	$L2_1$ = -5.48 + 0.64 L2 $L3_1$ = 0.14 + 0.46 L3 $L3_2$ = -0.98 + 0.33 L3
2 - 3 2 - 3 - 1 2 - 3 - 4	$L2_1$ = -3.05 + 0.61 L2 $L3_1$ = 1.68 + 0.43 L3 $L3_2$ = -3.00 + 0.34 L3 $L4_1$ = 26.35 + 0.34 L4 $L4_2$ = -20.88 + 0.38 L4 $L4_3$ = -17.66 + 0.27 L4
3 - 2 3 - 2 - 1	$L2_1$ = -2.33 + 0.63 L2 $L3_1$ = 5.62 + 0.45 L3 $L3_2$ = 1.25 + 0.31 L3

L21 = First length of two growing periods per year
L31 = First length of three growing periods per year
L32 = Second length of three growing periods per year
L41 = First length of four growing periods per year
L42 = Second length of four growing periods per year
L43 = Third length of four growing periods per year

In the climatic inventory map of Kenya, only the mean total dominant length has been inventoried on the map. The relationships in Table 3.6 are further presented in terms of length of growing period zones in Technical Annex 1; giving the mean total dominant (mapped) and the corresponding mean total associated (unmapped) lengths of growing periods. Similarly the relationships in Table 3.7 are further presented in terms of length of growing period zones in Technical Annex 1; giving the mean total length of growing period zones and the corresponding individual component lengths of growing periods.

3.1.2.3 Variability of Length of Growing Period

In addition to the frequency distribution mentioned in Section 3.1.2.1, coefficient of variation was calculated to allow a comparison of the variability in the mean length of growing period, and to take into account the likely losses in production. An aggregate relationship is given as follows.

Mean length of growing period (days)	Coefficient of variation (%)
< 30	> 50
30 - 59	50
60 - 89	45
90 - 119	40
120 - 149	35
150 - 179	30
180 - 209	25
210 - 239	20
240 - 269	15
270 - 299	10
> 299	< 10

3.1.2.4 Intermediate Lengths of Growing Periods

From the frequency distribution information (Section 3.1.2.1) occurrence of intermediate lengths of growing periods was quantified by relating P/ET ratio and moisture excess values with length of growing period.

The P/ET ratio for the intermediate lengths of growing periods of less than 150 days corresponds to values in the range 0.70 - 0.75.

The relationship which exists between the individual length of growing period and occurrence of intermediate periods is shown here.

Mean length of growing period (days)	Occurrence of intermediate periods (%)
< 30	100
30 - 59	65
60 - 89	25
90 - 119	10
120 - 149	5
> 149	< 1

3.1.3 Thermal Zones

To identify thermal zones, temperature criteria corresponding to the requirements of crops (including pasture and fuelwood), were taken into account (Technical Annexes 3, 4, 5, and 6)[1].

To cater for differences in temperature requirements of crops in the compilation of the country inventory, commensurate with the scale of the assessment (1:1 million), thermal regimes have been defined based on 2.5 °C intervals. A thermal difference of 2.5 °C corresponds to an altitudinal change of some 385 m, thus allowing a sufficiently fine matching of crop thermal requirements to prevailing thermal conditions as inventoried[2].

---

1 Field crops, pasture and fodder grasses and legumes, and fuelwood tree species have been classified into temperature-photosynthesis adaptability groups. Four temperature adaptability groups are recognized for field crops, four for pasture and fodder grasses and legumes and two for fuelwood species (each with three productivity classes).

2 Data from meteorological stations above 150 m altitude conform closely to the following relationship between average daily temperature in degree Celcius (T) and altitude in metres (A): $T = 30.2 - 6.496 \; (A / 1000)$. Temperature seasonality effects of latitude are minor due to the equatorial position of Kenya.

For Kenya nine reference thermal zones have been recognized as shown here.

Thermal zone code	Mean daily temperature range (°C)	Altitude (m)
1	> 25.0	< 800
2	22.5 - 25.0	800 - 1200
3	20.0 - 22.5	1200 - 1550
4	17.5 - 20.0	1550 - 1950
5	15.0 - 17.5	1950 - 2350
6	12.5 - 15.0	2350 - 2700
7	10.0 - 12.5	2700 - 3100
8	5.0 - 10.0	3100 - 3900
9	< 5.0	> 3900

3.1.4 Area Inventory of Climatic Resources

The area inventory of thermal zones, pattern of growing period zones and length of growing period zones, by district was prepared at 1 : 1 million scale.

The mapped inventory was compiled by:

(i) plotting the individual station data of temperature, pattern of length of growing period and mean total dominant length of growing period[1]; and

(ii) constructing boundaries of thermal zones, pattern of number of length of growing period zones, growing period zones and isolines of mean total dominant length of growing period with the values 0, 30, 60, 90, 120, 150, 180, 210, 240, 270, 300, 330, 365⁻ and 365⁺ days respectively, delineating the mean total dominant length of growing period zones of 0, 1-29, 30-59, 60-89, 90-119 120-149, 150-179, 180-209, 210-239, 240-269, 270-299, 300-329, 330-364, 365⁻ and 365⁺ days.

In addition to normal extrapolation techniques, extensive use was made of landsat images, climatic maps, vegetation maps, land use maps, topographic maps and soil maps to guide the delineation of boundaries and isolines.

The three climatic inventories in map form were digitized, and the digitized information from the maps was converted to a grid cell data base.

A generalized map thermal zones is presented in Figure 3.5, and their extents are presented in Table 3.8. A generalized map of mean total dominant length of growing period zones is presented in Figure 3.6, and their extents are presented in Table 3.9. A generalized map of pattern of length of growing period zones is presented in Figure 3.7, and their extents are presented in Table 3.10.

Extents of mean total dominant length of growing period zones by pattern of length of growing period zones for each of the thermal zones, and for all thermal zones combined, are presented in Technical Annex 1.

3.2 Soil Resources

3.2.1 Exploratory Soil Map of Kenya

The Exploratory Soil Map of Kenya (Siderius and van der Pouw 1980; Sombroek, Braun and van der Pouw 1982) at 1:1 million scale was used to compile the soil resources inventory for this assessment.

---

1 The individual component lengths of both the dominant (mapped) and associated length of growing period zones are presented in Technical Annex 1.

FIGURE 3.5
Generalized map of thermal zones

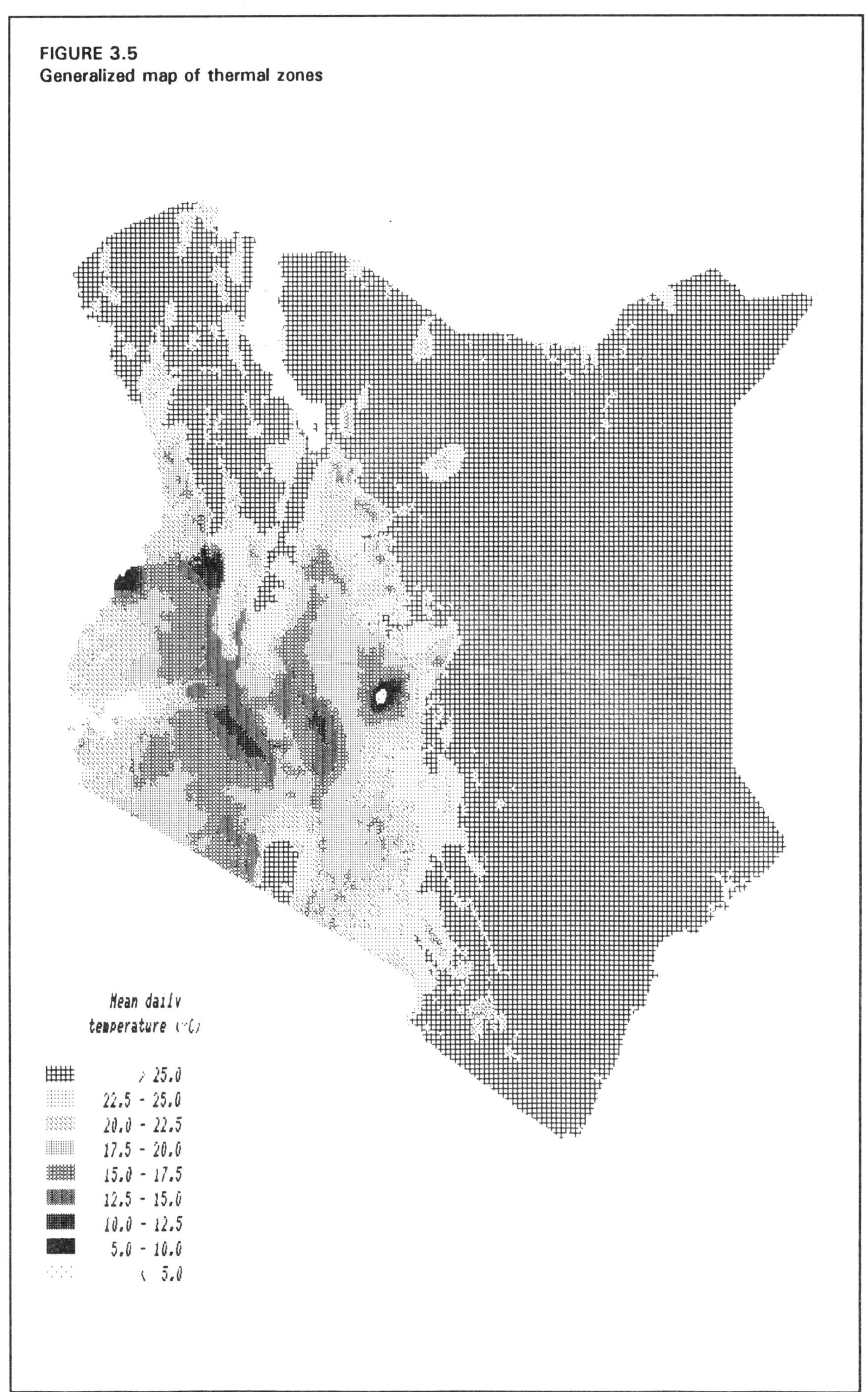

FIGURE 3.6
Generalized map of mean total dominant length of growing period zones

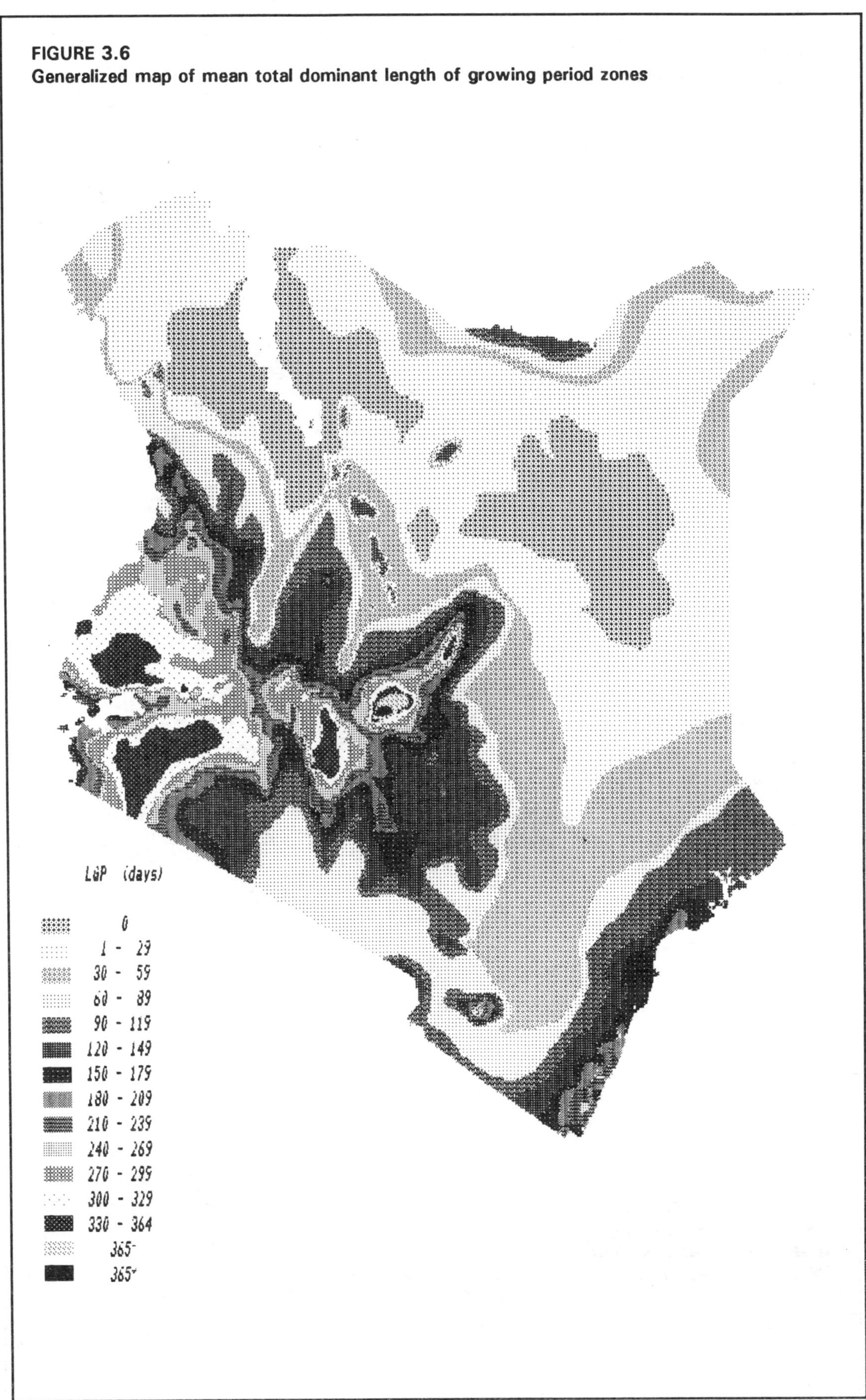

FIGURE 3.7
Generalized map of pattern of length of growing period zones

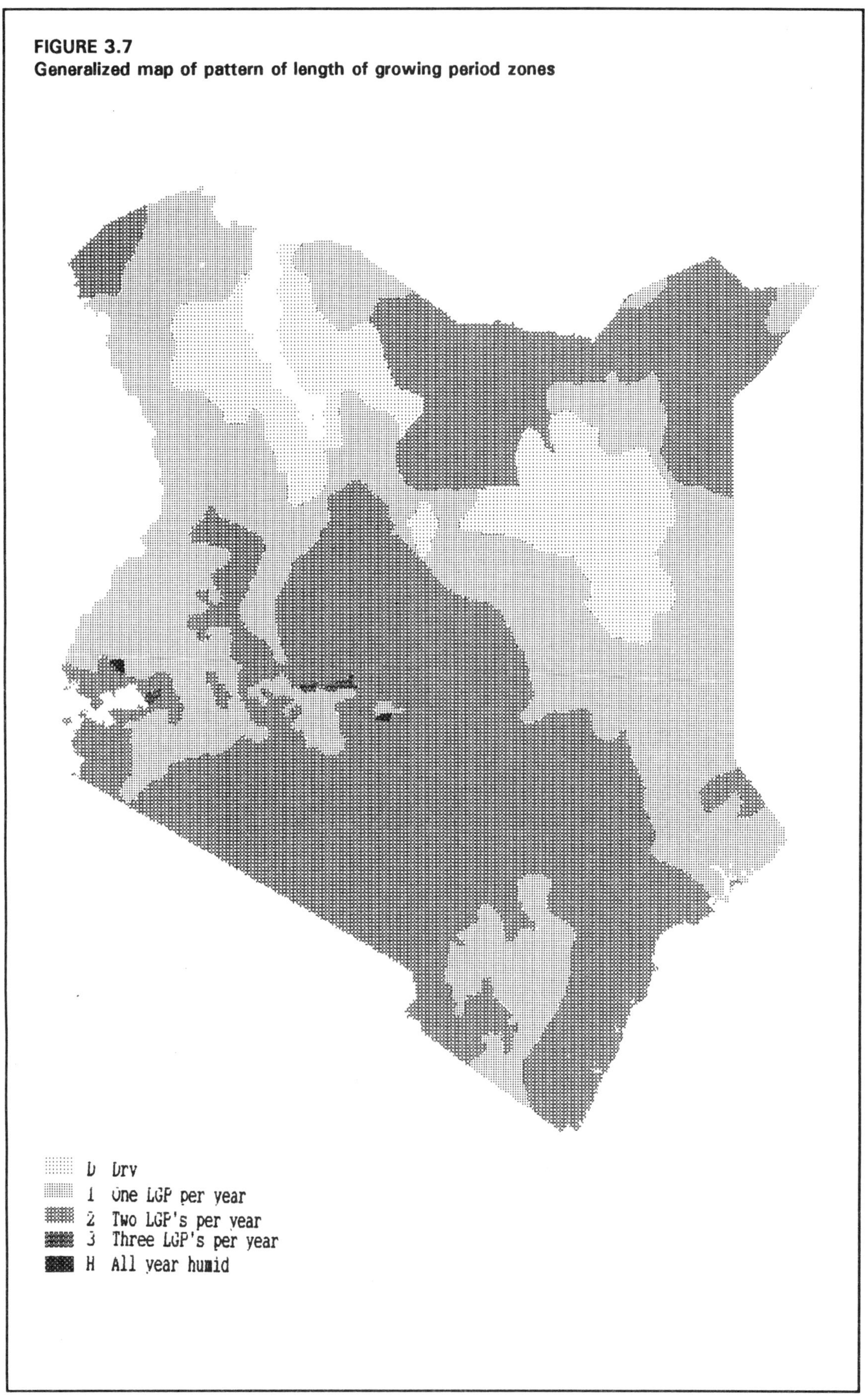

TABLE 3.8
Extents of thermal zones

Thermal zone code	Mean daily temperature range (°C)	Extent (ha)	Percentage of total area
1	> 25.0	38 120 600	66.17
2	22.5 - 25.0	5 783 600	10.04
3	20.0 - 22.5	4 070 100	7.07
4	17.5 - 20.0	4 484 400	7.78
5	15.0 - 17.5	3 448 300	5.99
6	12.5 - 15.0	1 268 200	2.20
7	10.0 - 12.5	307 100	0.53
8	5.0 - 10.0	107 400	0.19
9	< 5.0	17 500	0.03
1 - 9		57 607 200	100.00

TABLE 3.9
Extents of mean total dominant length of growing period zones

LGP zone code	LGP zone (days)	Extent (ha)	Percentage of total area
1	0	6 837 700	11.87
2	1 - 29	14 941 300	25.94
3	30 - 59	9 544 300	16.57
4	60 - 89	6 027 100	10.46
5	90 - 119	5 019 100	8.71
6	120 - 149	4 437 100	7.70
7	150 - 179	2 377 900	4.13
8	180 - 209	1 380 500	2.40
9	210 - 239	1 204 800	2.09
10	240 - 269	1 192 800	2.07
11	270 - 299	1 711 600	2.97
12	300 - 329	1 627 200	2.82
13	330 - 364	1 223 300	2.12
14	365^-	57 300	0.10
15	365^+	25 200	0.04
1 - 15		57 607 200	100.00

This soil map, published by Kenya Soil Survey in 1980, provides the latest country-wide soil data base and includes information on distribution and characteristics of soils, landform and geology/parent material.

3.2.2 Soil Mapping Units

The soil mapping units are soil associations or soil complexes composed of dominant soils, associated soils and inclusions. The soil mapping units have been registered on the map by a symbol reflecting the landform in which they occur. In Table 3.11 the occurrences of soil mapping units are presented by landform.

The Exploratory Soil Map consists of 390 different soil mapping units. Extents of the individual soil mapping units are presented in Technical Annex 1.

TABLE 3.10
Extents of pattern of length of growing period zones

Pattern zone code	Pattern zone symbol	Extent (ha)	Percentage of total area
1	1	30 800	0.05
2	H - 1	25 200	0.04
3	1 - H	330 400	0.57
4	1 - H - 2	611 200	1.07
5	1 - 2 - H	52 300	0.09
6	1 - 2	3 687 300	6.40
7	1 - 2 - 3	615 200	1.07
8	1 - 3 - 2	17 700	0.03
9	1 - 2 - D	3 488 000	6.05
10	1 - D - 2	6 080 700	10.56
11	1 - D	6 218 900	10.80
12	2	533 800	0.93
13	2 - 1	20 332 500	35.30
14	2 - 1 - H	42 900	0.07
15	2 - 1 - 3	3 326 800	5.77
16	2 - 3	1 633 800	2.84
17	2 - 3 - 1	2 696 100	4.68
18	2 - 3 - 4	53 900	0.09
19	2 - 1 - D	698 700	1.21
20	3 - 2	76 500	0.13
21	3 - 2 - 1	219 100	0.38
22	D	6 835 400	11.87
1 - 22		57 607 200	100.00

For each soil mapping unit the following semi-quantified information in terms of description, classes and extents has been transferred to the soil resources data base of this assessment:

- Landform
- Geology/Parent material
- Soil units (with implied characteristics)
- Slope-gradient classes
- Soil texture classes
- Soil phases

3.2.3 Landform

Landform is the first entry in the legend of the Exploratory Soil Map. It provides information on physiography, altitudinal position and slope patterns. A generalized map of landforms in Kenya is presented in Figure 3.8, and a description of landforms and their extents is presented in Table 3.11.

For the Exploratory Soil Map of Kenya six slope classes have been employed in 12 combinations. The slope classes are: A, 0-2%; B, 2-5%; C, 5-8%; D, 8-16%; E 16-30%; and F > 30%.

TABLE 3.11
Occurrence of soil mapping units by landform

Landform symbol	Landform description	Soil Mapping units	Extent (000 ha)	Percentage of total area
A	Floodplain	1 - A18 A8 + A12	3 179	5.52
B	Bottomland	B1 - B16	973	1.69
D	Dunes or dune land	D1 - D3 D1 + P13	90	0.16
F	Footslopes	F1 - F19	2 111	3.66
FY	Footslopes and piedmont plain (undifferentiated)	FY1 - FY3	614	1.07
H	Hills and minor scarps	H1 - H22	3 204	5.56
Hs	Step-faulted scarp of the Rift Valley	Hs1	515	0.89
L	Plateau and high level structural plain	L1 - L31	4 052	7.03
La	Lava flow	Lava	960	1.67
Lc	Coastal plateau	Lc1 - Lc31	213	0.37
Ls	Step-faulted floor of the Rift Valley	Ls1 - Ls3	811	1.41
Lu	Plateau/upper-level upland transition	Lu1 - Lu2	98	0.17
M	Mountains and major scarps	M1 - M12	2 375	4.12
Pch	Higher-level coastal plain	Pc1 - Pc3	430	0.75
Pcl	Lower-level coastal plain	Pc4 - Pc7	599	1.04
Pcr	Reef coastal plain	Pc8 - Pc10	83	0.14
Pd	Dissected erosional plain	Pd1 - Pd6	1 895	3.29
Pf1	Sedimentary plain of large alluvial plains (older fans)	Pf1 - Pf3	312	0.54
Pf2	Sedimentary plain of large alluvial plains (younger fans)	Pf4 - Pf5	377	0.65
Pl	Lacustrine plain	Pl1 - Pl13	863	1.50
Pn	Non-dissected erosional plain	Pn1 - Pn35	6 007	10.43
Psh	Higher-level sedimentary plain	Ps1 - Ps6 Ps3 + Ps15	3 885	6.74
Psl	Lower-level sedimentary plain	Ps21 - Ps27	3 138	5.54
Psm	Middle sedimentary plain ('enclosed') plain and sealing loam plain)	Ps7 - Ps20 Ps11 + D1	5 618	9.75
Psx	Sedimentary plain of undifferentiated level	Ps28 - Ps29 Ps28 + D1	247	0.43
Pt	Sedimentary plain of upper river terrace	Pt1 - Pt4	323	0.56
Pv	Volcanic plain	Pv1 - Pv12	998	1.73
R	Volcanic footridges	R1 - R14	3 121	5.42
S	Swamp	S1 - S3	95	0.16
T	Mangrove swamp	T	134	0.23
Uc	Coastal upland	Uc1 - Uc11	533	0.93
Uh	Upper middle-level upland	Uh1 - Uh19	786	1.36
Ul	Lower-level upland	Ul1 - Ul21	1 400	2.43
Um	Lower middle-level upland	Um1 - Um29 Up1	2 090	3.63
Up	Upland/high-level plain transitional land	- Up8	386	0.67
Uu	Upper-level upland	Uu1 - Uu3	132	0.23
Ux	Upland, undifferentiated land	Ux1 - Ux10	1 744	3.03
V	Minor valley	V1 - V2	112	0.20
W	Badland	W1 - W2	722	1.25
Y	Piedmont plain	Y1 - Y13	2 134	3.70
Z1	Older coastal beach ridge	Z1	79	0.14
Z2	Younger coastal beach ridge	Z2	39	0.07
Z3	Lakeside beach ridge	Z3	5	0.01
Lakes			118	0.20
Towns			11	0.02
Total extent			57 607	100.00

A generalized map of slope-gradient classes in Kenya is presented in Figure 3.9. The combinations of slope classes employed and their extents are presented in Table 3.12.

To each of the 12 slope classes, inventoried in the Exploratory Soil Map, associated slope classes have been assigned. These associated slope classes, covering up to 10% of the

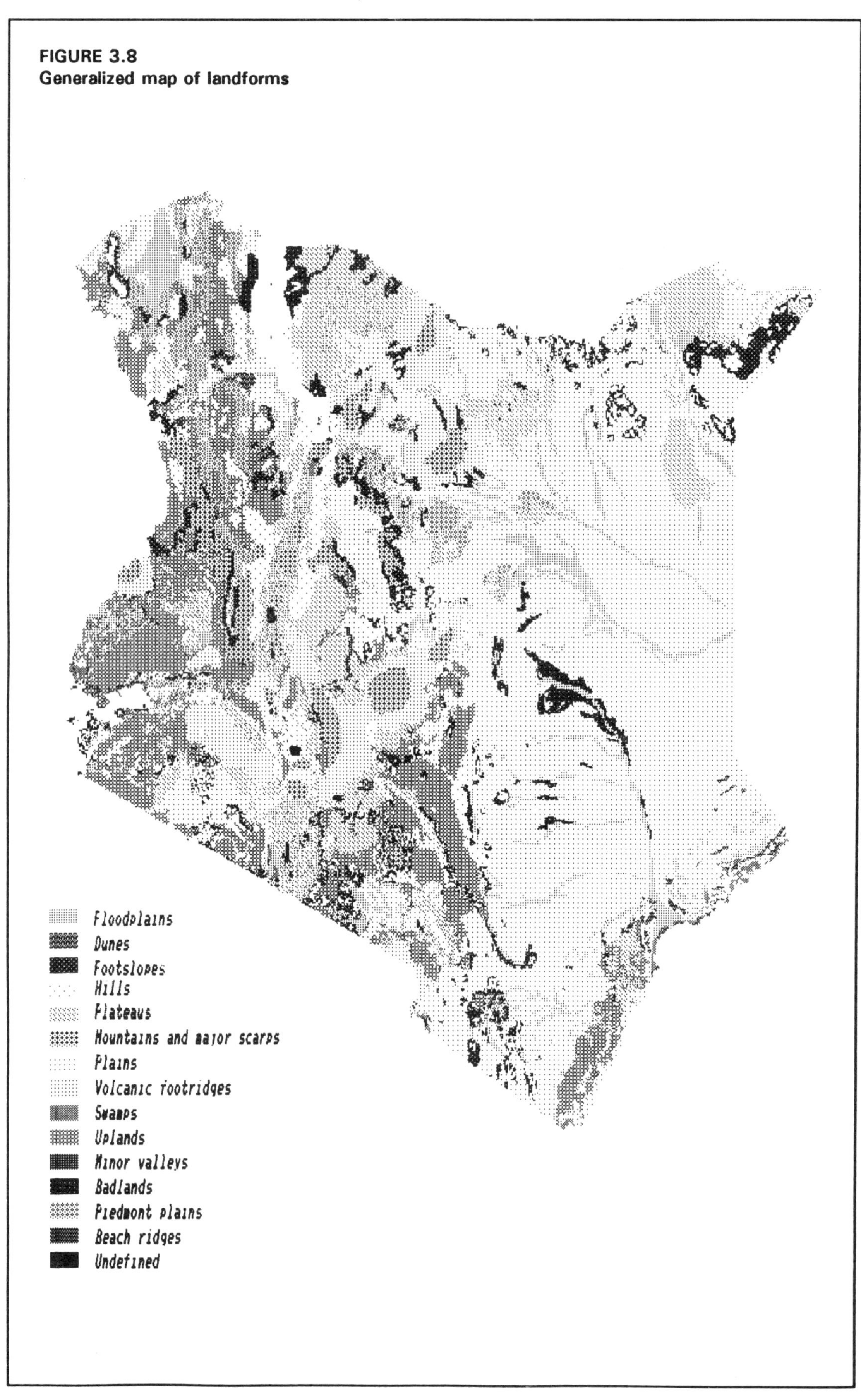

FIGURE 3.8
Generalized map of landforms

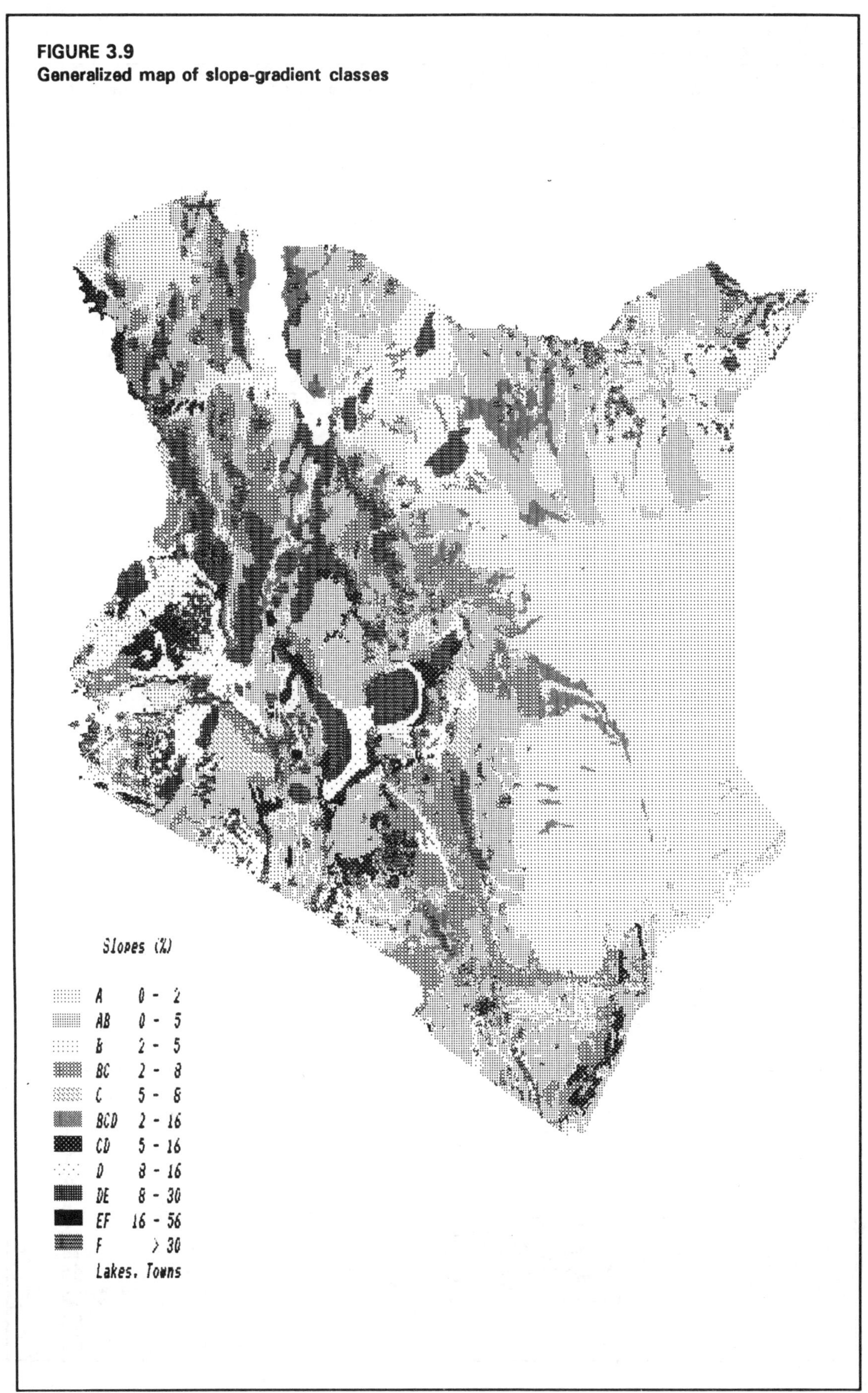

FIGURE 3.9
Generalized map of slope-gradient classes

TABLE 3.12
Extents of slope classes

Slope class symbol	Slope class (%)	Extent (ha)	Percentage of total area
A	0 - 2	19 868 850	34.49
AB	0 - 5	14 540 600	25.24
B	2 - 5	3 098 150	5.38
BC	2 - 8	7 351 200	12.76
C	5 - 8	759 200	1.32
BCD	2 - 16	1 707 300	2.96
CD	5 - 16	1 557 450	2.70
D	8 - 16	1 616 200	2.81
DE	8 - 30	885 850	1.54
E, EF, F	> 16	6 093 400	10.58
Lakes		117 600	0.20
Towns		10 700	0.02
Total extent		57 607 200	100.00

TABLE 3.13
Associated slope classes

Slope class symbol	%	Associated slope classes					
A	0 - 2	100%	A				
AB	0 - 5	100%	AB				
B	2 - 5	100%	B				
BC	2 - 8	90%	BC	5%	A	5%	D
C	5 - 8	90%	C	5%	AB	5%	D
BCD	2 - 16	90%	BCD	5%	A	5%	E
CD	5 - 16	90%	CD	5%	AB	5%	E
D	8 - 16	90%	D	5%	BC	5%	E
DE	8 - 30	90%	DE	5%	BC	5%	F
E	16 - 30	90%	E	5%	BCD	5%	F
EF	16 - 56	95%	EF	5%	BCD		
F	30 - 56	95%	F	5%	DE		

TABLE 3.14
Quartiles of slope classes

Slope class symbol	%	Gentlest Q1	Lower Q2	Upper Q3	Steepest Q4
A	0 - 2	0	1	1	2
AB	0 - 5	0	2	4	5
B	2 - 5	2	3	4	5
BC	2 - 8	2	4	6	8
C	5 - 8	5	6	7	8
BCD	2 - 16	2	6	11	16
CD	5 - 16	5	9	12	16
D	8 - 16	8	11	13	16
DE	8 - 30	8	16	22	30
E	16 - 30	16	21	25	30
EF	16 - 56	16	30	42	56
F	30 - 56	30	39	47	56

land area of the 12 slope classes, are used for evaluation purposes and included in the land resources inventory. The inventoried slope classes and associated slope classes are presented in Table 3.13. For the same purposes assumed mean slopes of quartiles of the land area of each of the slope classes have been assigned. These values are presented in Table 3.14.

3.2.4 Geology

Geology is the second entry in the legend of the Exploratory Soil Map. The geological subdivisions reflect mainly resistance to weathering and richness of parent material in order to provide linkage with soil formation. The first level subdivision comprises three types of rocks: igneous rocks, metamorphic rocks and sedimentary rocks.

Igneous rocks and metamorphic rocks are futher subdivided from basic to acid. The sedimentary rocks are futher subdivided from fine to coarse textured.

Each soil mapping unit is accordingly characterized for its geological setting/parent material. A generalized map of geology/parent material is presented in Figure 3.10 and the descriptions and extents of geological units are presented in Table 3.15.

3.2.5 Soil Units

The individual soil units of the soil associations or soil complexes (soil mapping units of the Exploratory Soil Map) have been defined in accordance with the FAO-Unesco legend of the Soil Map of the World (FAO 1974). The soil units adopted were selected on the basis of present knowledge on the formation, characteristics and distribution of the soils, their importance as resources for agricultural production and their significance as a factor of the environment.

In the legend of the Exploratory Soil Map some adaptations of the FAO-Unesco legend have been introduced. At the first level (great group) the terminology for Lithosols and Nitosols has been modified. At the second level (unit level) new subgroups have been introduced (cambic and orthic Rendzinas) and others modified (vertic Gleysols, mollic Nitisols, chromic Acrisols, chromic Luvisols and chromic Cambisols). In order to reflect the greater amount of detail of the Exploratory Soil Map of Kenya, a third level of terminology (subunit level) has been introduced for subdivision of soil units into subunits. The prefixes used to distinguish subunits are ando, calcaro, chromo, ferralo, luvo, nito, ortho and verto.

In the Exploratory Soil Map, 123 different soil units and five miscellaneous units occur. Table 3.16 presents the soil units and miscellaneous units and their extents.

The soil units have been defined in terms of measurable and observable properties of the soil itself, and specific clusters of such properties are combined into 'diagnostic horizons' and 'diagnostic properties'.

The diagnostic horizons have been used as defined in the FAO-Unesco legend. Diagnostic horizons and properties of the soil units are given in Technical Annex 1.

Complete definitions of soil units are given in Volume 1 (Legend) of FAO-Unesco Soil Map of the World (FAO 1974) and Exploratory Soil Survey Report (KSS 1982a).

FIGURE 3.10
Generalized map of geology/parent material

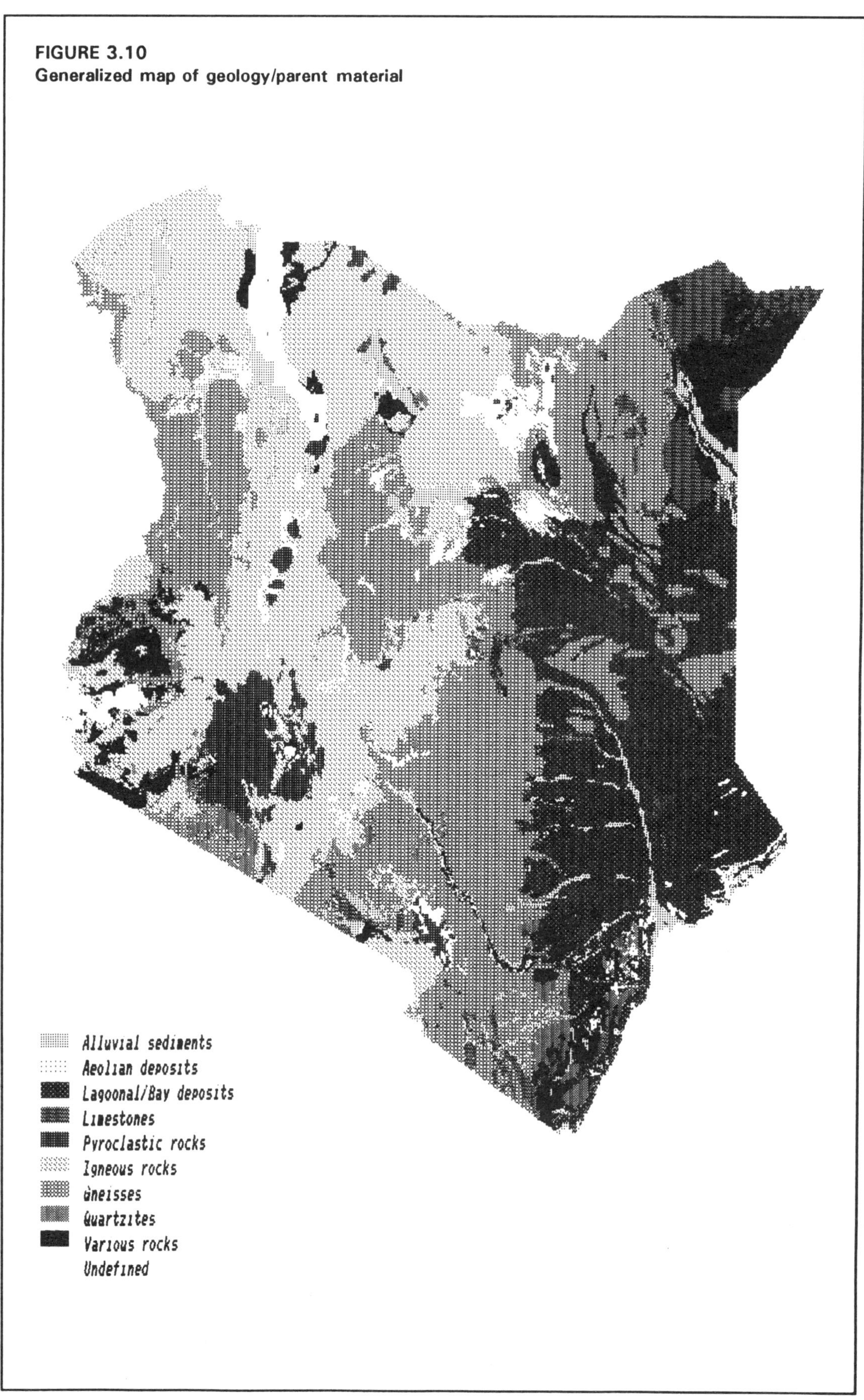

TABLE 3.15
Geology/parent material

Geology symbol	Geology description	Extent (ha)	Percentage of total area
A	(Alluvial) Sediments from various sources[1]	2 241 600	3.89
B	Basic and ultra-basic igneous rocks	6 786 800	11.78
B^+	As in B, but with volcanic ash admixture	233 300	0.39
BP	As in B, but volcanic ash predominant	52 200	0.09
D	Mudstones, claystones	102 900	0.18
E	Aeolian sediments (cover sands)	235 400	0.41
F	Gneisses rich in ferromagnesian minerals, hornblende gneisses	1 680 600	2.92
G	Granites, granodiorites	477 900	0.83
G^+	As in G, but with volcanic ash admixture	14 400	0.02
GF	Biotite-hornblende granites	38 500	0.07
GF^+	As in GF, but with volcanic ash admixture	70 700	0.12
GP	As in G, but volcanic ash predominant	19 300	0.03
GR	Complex of G and R	50 700	0.09
I	Intermediate igneous rocks (syenites etc.)	499 900	0.87
I^+	As in I, but with volcanic ash admixture	90 900	0.16
J	Lagoonal deposits	1 154 600	2.00
K	Siltstones	1 466 600	2.55
KT	Complex of K and T	94 100	0.16
L	Limestones, calcitic mudstones	2 056 400	3.57
N	Biotite gneisses	400 600	0.70
N^+	As in N, but with volcanic ash admixture	160 300	0.28
O	Plio-pleistocene bay sediments	9 606 600	16.28
P	Pyroclastic rocks	1 897 400	3.29
Q	Quartzites	405 900	0.70
R	Quartz-feldspar gneisses	58 900	0.10
S	Feldspar, grits, arkoses	647 400	1.12
T	Shales	164 000	0.28
U	Undifferentiated basement system rocks	14 007 000	24.31
U^+	As in U, but with volcanic ash admixture	1 172 400	2.04
UP	As in U, but volcanic ash predominant	36 700	0.06
V	Undifferentiated or various igneous rocks	8 536 300	14.28
W	Marls	168 500	0.29
X	Undifferentiated or various rocks	1 350 200	2.34
X^+	As in X, but with volcanic ash admixture	11 700	0.02
Y	Acid igneous rocks (rhyolite, aplite)	155 800	0.27
Y^+	As in Y, but with volcanic ash admixture	79 000	0.14
-	Not defined	1 263 400	2.19
Lakes		117 600	0.20
Towns		10 700	0.02
Total extent		56 607 200	100.00

1 If the source of alluvial sediments and bottomland infills is known (e.g., basalts), then the code for this rock is used, otherwise the code A applies.

TABLE 3.16
Extents of soil units

Symbol	Soil unit name	Extent (ha)	% total area
A	Acrisols	45 080	0.08
Ac	Chromic Acrisols	82 830	0.14
Ag	Gleyic Acrisols	49 950	0.09
Ah	Humic Acrisols	151 345	0.26
Aic	Ferralo-chromic Acrisols	772 685	1.34
Aif	Ferralo-ferric Acrisols	127 760	0.22
Aio	Ferralo-orthic Acrisols	325 475	0.56
Ao	Orthic Acrisols	167 920	0.29
Ap	Plinthic Acrisols	14 880	0.03
Ath	Ando-humic Acrisols	14 600	0.03
B	Cambisols	8 800	0.02
Bc	Chromic Cambisols	1 000 700	1.74
Bd	Dystric Cambisols	70 350	0.12
Be	Eutric Cambisols	630 140	1.09
Bf	Ferralic Cambisols	53 840	0.09
Bg	Gleyic Cambisols	31 150	0.05
Bh	Humic Cambisols	513 185	0.89
Bk	Calcic Cambisols	1 890 630	3.28
Bnc	Nito-chromic Cambisols	296 195	0.51
Btc	Ando-chromic Cambisols	436 600	0.76
Bte	Ando-eutric Cambisols	181 750	0.32
Bv	Vertic Cambisols	20 825	0.04
Ch	Haplic Chernozems	39 800	0.07
Ck	Calcic Chernozems	56 800	0.10
Ec	Cambic Renzinas	14 700	0.03
Eo	Orthic Renzinas	184 790	0.32
F	Ferralsols	128 080	0.22
Fa	Acric Ferralsols	59 400	0.10
Fh	Humic Acrisols	6 100	0.01
Fnh	Nito-humic Ferralsols	13 600	0.02
Fnr	Nito-rodic Ferralsols	225 900	0.39
Fo	Orthic Ferralsols	855 490	1.49
Fr	Rodic Ferralsols	2 695 015	4.68
Fx	Xanthic Ferralsols	73 860	0.13
G/Ge	Gleysols/Eutric Gleysols	9 270	0.02
Gc	Calcaric Gleysols	95 100	0.17
Gd	Dystric Gleysols	22 610	0.04
Gh	Humic Gleysols	52 830	0.09
Gm	Mollic Gleysols	124 840	0.22
Gv	Vertic Gleysols	820 070	1.42
Hg	Gleyic Phaeozems	147 180	0.26
Hh	Haplic Phaeozems	96 355	0.17
Hnl	Nito-luvic Phaeozems	21 900	0.04
Hol	Orthic-luvic Phaeozems	319 120	0.55
Hrl	Chromo-luvic Phaeozems	523 270	0.91
Hth	Ando-haplic Phaeozems	36 190	0.06
Htl	Ando-luvic Phaeozems	263 980	0.46
Hvl	Verto-luvic Phaeozems	642 225	1.11
I	Lithosols	2 344 045	4.07
Ir	Ironstone soils	216 285	0.38
J	Fluvisols	34 800	0.06
Jc	Calcaric Fluvisols	1 434 600	2.49
Je	Eutric Fluvisols	299 400	0.52
Jt	Thionic Fluvisols	79 650	0.14
Kh	Haplic Kastanozems	57 640	0.10
L	Luvisols	194 750	0.34
La	Albic Luvisols	124 625	0.22
Lc	Chromic Luvisols	2 321 880	4.03

TABLE 3.16 (Continued)

Symbol	Soil unit name	Extent (ha)	% total area
Lf	Ferric Luvisols	582 150	1.01
Lg	Gleyic Luvisols	167 395	0.29
Lic	Ferralo-chromic Luvisols	784 840	1.37
Lif	Ferralo-ferric Luvisols	44 955	0.08
Lio	Ferralo-orthic Luvisols	50 325	0.09
Lk	Calcic Luvisols	981 995	1.70
Lnc	Nito-chromic Luvisols	34 120	0.06
Lnf	Nito-ferric Luvisols	36 260	1.81
Lo	Orthic Luvisols	1 039 855	0.29
Lv	Vertic Luvisols	169 710	0.12
Mo	Orthic Greyzems	70 700	0.07
Mvo	Verto-orthic Greyzems	38 100	0.05
Nd	Dystric Nitisols	31 195	0.05
Ne	Eutric Nitisols	600 610	1.04
Nh	Humic Nitisols	635 860	1.10
Nm	Mollic Nitisols	190 610	0.33
Nth	Ando-humic Nitisols	212 530	0.37
Nve	Verto-eutric Nitisols	33 365	0.06
Nvm	Verto-mollic Nitisols	3 240	0.01
Od	Distric Histosols	79 800	0.14
Q	Arenosols	15 750	0.03
Qa	Albic Arenosols	63 880	0.11
Qc	Cambic Arenosols	702 790	1.22
Qf	Ferralic Arenosols	2 200 890	3.82
Qkc	Calcaro-cambic Arenosols	446 780	0.78
Ql	Luvic Arenosols	187 860	0.33
R	Regosols	13 250	0.02
Rc	Calcaric Regosols	1 256 110	2.18
Rd	Dystric Regosols	138 640	0.24
Re	Eutric Regosols	696 390	1.21
Rtc	Ando-calcaric Regosols	301 440	0.52
S	Solonetz	634 580	1.10
Sg	Gleyic Solonetz	434 030	0.75
Slo	Luvo-orthic Solonetz	4 664 100	8.10
Sm	Mollic Solonetz	43 820	0.08
So	Orthic Solonetz	2 536 730	4,40
Th	Humic Andosols	455 520	0.79
Tm	Mollic Andosols	528 950	0.92
Tv	Vitric Andosols	81 600	0.14
U	Rankers	154 670	0.27
V	Vertisols	149 580	0.26
Vc	Chromic Vertisols	763 920	1.33
Vp	Pellic Vertisols	1 514 390	2.63
W	Planosols	4 380	0.01
Wd	Dystric Planosols	114 205	0.20
Wh	Eutric Planosols	323 710	0.56
We	Humic Planosols	195 800	0.34
Ws	Solodic Planosols	2 472 405	4.29
Wve	Verto-eutric Planosols	251 205	0.44
X	Xerosols/Yermosols	102 920	0.18
Xh	Haplic Xerosols/Yermosols	650 620	1.13
Xk	Calcic Xerosols/Yermosols	3 620 670	6.29
Xy	Gypsic Xerosols/Yermosols	168 500	0.29
Z	Solonchaks	577 750	1.00
Zg	Gleyic Solonchaks	187 000	0.32
Zo	Orthic Solonchaks	1 601 410	2.78
Zt	Takyric Solonchaks	147 800	0.26

TABLE 3.16 (Continued)

Symbol	Soil unit name	Extent (ha)	% total area
Lava		1 015 960	1.76
Lava flows		128 990	0.22
Lava fields		56 460	0.10
Rock outcrops		557 885	0.97
Ice cap		5 335	0.01
Lakes		117 600	0.20
Towns		10 700	0.02
Total extent		57 607 200	100.00

3.2.6 Soil Textures

Soil textures may vary within the range of textures defined for a particular soil unit. In the legend of the Exploratory Soil Map textural classes for individual soil units by soil mapping unit are presented. The three major textural divisions are subdived into 17 classes. The three major textural divisions are shown in the texture diagram.

Textural classes by major division are also listed below; percentage occurrence is presented in Table 3.17.

Coarse textures
- Sand
- Loamy coarse sand
- Fine sand
- Loamy fine sand
- Loamy sand

Medium textures
- Fine sandy loam
- Sandy loam
- Loam
- Sandy clay loam
- Silt loam
- Clay loam
- Silty clay loam
- Silt

Fine textures
- Sandy clay
- Silty clay
- Peaty clay

TABLE 3.17
Percentage occurrence of textural classes

Texture class	Texture symbol	Percentage [1]
Sand	S	0.2
Loamy coarse sand	LCS	0.5
Fine sand	FS	0.2
Loamy fine sand	LFS	0.8
Loamy sand	LS	3.8
Σ Coarse textures		5.5
Fine sandy loam	FSL	1.3
Sandy loam	SL	3.1
Loam	L	6.4
Sandy clay loam	SCL	15.4
Silt loam	SL	0.6
Clay loam	CL	28.5
Silty clay loam	SICL	0.4
Silt	SI	<0.1
Σ Medium textures		55.8
Sandy clay	SC	10.2
Silty clay	SIC	0.3
Peaty clay	PC	<0.1
Clay	C	28.2
Σ Fine textures		38.7

1 Total extent where texture is applicable is 55 714 270 ha (96.9% of total area).

3.2.7 Soil Stoniness

The presence of coarse material (stoniness) in the soil profile has been inventoried separately from soil textures. The presence of coarse material is subdivided into six types:

Gravely (G)

Very gravely	(VG)
Stony	(S)
Bouldery	(SB)
Stony/bouldery	(SB)
Bouldery/stony	(BS)

3.2.8 Soil Phases

Soil phases indicate land characteristics which are not considered in the definition of the soil units but are significant to the use and management of land. The soil phases recognized on the Exploratory Soil Map of Kenya can be grouped into phases indicating a mechanical hindrance or limitation (rocky, bouldery, boulder-mantle, stony, stone-mantle, gravel-mantle); phases indicating an effective soil depth limitation (lithic, paralithic, petro-calcic, piso-calcic, petro-ferric, piso-ferric); and phases indicating a physico-chemical limitation (saline, sodic and saline-sodic). Soil phases can occur alone (one soil phase or in combination (two or three phases). Some of these phases have been defined in the FAO-Unesco Legend, while others have been introduced in the Exploratory Soil Map. The soil phases and their extents are presented in Table 3.18.

3.2.9 Soil Mapping Unit Composition

At the exploratory level a soil mapping unit only rarely comprises a single soil; usually it consist of one main soil with minor associates. When the various soils of a soil mapping unit occur in a recognizable geographical pattern in defined proportions, they constitute a soil association; if such a pattern is absent, they form a soil complex.

The productivity potential of different soil units within a soil mapping unit consequently may vary widely. The suitability of soil association (soil complex) for specific use cannot be assessed without taking account of each individual soil unit within the association.

The legend of the Exploratory Soil Map does not provide explicit information in a quantified manner on the composition of the soil mapping units. This however has been provided to FAO by Stiboka in Wageningen in close cooperation with Kenya Soil Survey in Nairobi and the International Soil Reference and Information Centre in Wageningen (van der Pouw 1983).

The complete mapping unit composition table is presented in Technical Annex 1. This table consists of the percentage allocation of soil units by slope class, soil texture and soil phases for each soil mapping unit, and of information on landform and geology/parent material.

The composition of a example soil association (Ps3) is illustrated and explained in Figure 3.11.

3.3 Land Use Overlays

The land use and tse-tse infestation overlays have been integrated into the physical resources inventories. Consequently each agro-ecological cell is further characterized by the information contained in these overlays.

TABLE 3.18
Extents (%) of soil phases

Soil phase	Symbol	Percentage[1]
Rocky	R	6.7
Bouldery	B	1.6
Boulder-mantle	BM	1.1
Stony	S	4.6
Stone-mantle	SM	5.1
Gravely	G	0.2
Gravel-mantle	GM	0.9
Lithic	P	3.7
Paralithic	PP	0.7
Petrocalcic	K	0.2
Petrocalcic	KK	3.1
Pisocalcic	C	0
Pisocalcic	CC	0.4
Petroferric	M	1.2
Pisoferric	N	0.1
Saline	A	12.8
Sodic	O	9.7
Saline-sodic	AO	10.6
Fragipan	F	0.1
One soil phase		62.8
Rocky and bouldery	R and B	3.7
Rocky and stony	R and S	2.9
Bouldery and stony	B and S	3.6
Boulder-mantle and saline-sodic	BM and AO	1.1
Stony and rocky	S and R	0.3
Stony and bouldery	S and B	<0.1
Stony and petrocalcic	S and K	<0.1
Stony and saline-sodic	S and AO	1.0
Stone-mantle and sodic	SM and O	1.0
Stone-mantle and saline-sodic	SM and AO	0.1
Lithic and rocky	P nd R	1.8
Lithic and bouldery	P and B	1.1
Lithic and boulder-mantle	P and BM	1.3
Lithic and stony	P and S	2.6
Lithic and sodic	P and O	<0.1
Lithic and saline-sodic	P and AO	0.1
Paralithic and rocky	PP and R	<0.1
Paralithic and stony	PP and S	0.7
Petrocalcic and stony	K and S	1.6
Petrocalcic and saline-sodic	K and AO	<0.1
Petrocalcic and saline	KK and A	0.1
Petrocalcic and sodic	KK and O	0.1
Pisoferric and rocky	N and R	0.1
Pisoferric and petroferric	N and M	0.2
Pisoferric and fragipan	A and F	0.1
Sodic and fragipan	O and F	0.1
Two soil phases		23.7
Rocky and bouldery and saline-sodic	R and B and AO	1.1
Rocky and lithic and stony	R and P and S	0.1
Bouldery and stony and saline	B and S and A	0.2
Bouldery and stony and saline-sodic	BM and S and AO	0.8
Lithic and rocky and bouldery	P and R and B	0.2
Lithic and rocky and stony	P and R and S	2.8
Lithic and bouldery and stony	P and B and S	0.2
Lithic and bouldery and saline	P and B and A	1.0
Lithic and boulder-mantle and saline-sodic	P and BM and AO	1.1
Lithic and stony and rocky	P and S and R	0.8
Lithic and stony and saline	P and S and A	0.6
Lithic and stony and saline-sodic	P and S and AO	0.1
Lithic and stone-mantle and saline-sodic	P and SM and AO	3.2
Lithic and gravel-mantle and saline	P and GM and S	1.3
Three soil phases		13.5

1 Total extent where soil phases occur is 28,096,674 ha (48.9 % of total area)

FIGURE 3.11
Example of soil mapping unit composition of soil mapping unit Ps3

Soil mapping unit Ps3 is an associaton of ferralo-chromic Luvisols (Lic), ferralic Arenosols (Qf) and ferric Luvisols (Lf)

Soil-unit	Soil-texture	Soil-phase	Slope	Proportion
Lic	Sandy clay loam	n.a.	0-2%	60%
Qf	Sandy loam	n.a.	0-2%	20%
Lf	Sandy clay loam	n.a.	0-2%	20%

(Total extent of all Ps3 units is 780 400 ha)

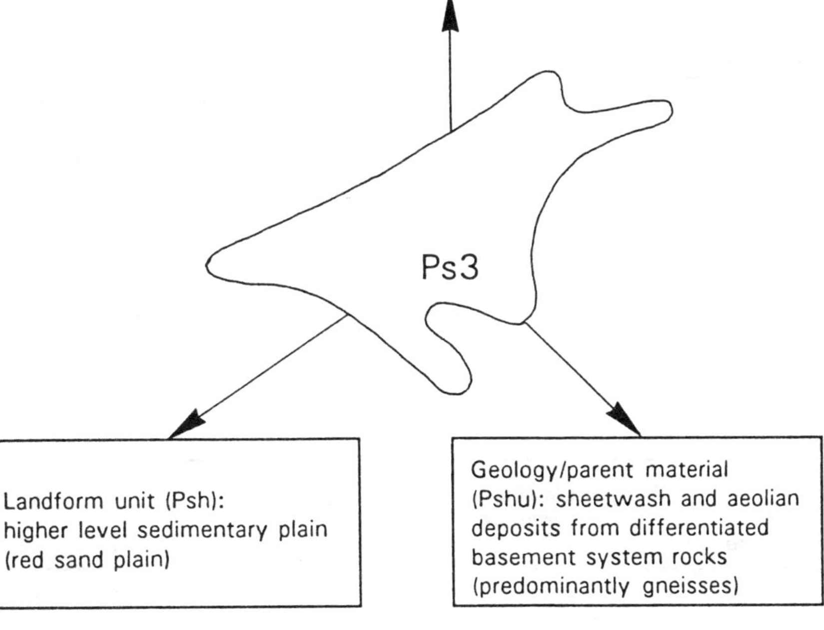

TABLE 3.19
Extents of cash crop zones

Crop symbol	Crop code	Crop description	Extent (ha)	Percentage area
01	2	Tea (secondary)	422 600	8.37
02	3	Coffee (secondary)	441 700	8.75
03	4	Sugarcane (secondary)	848 500	16.81
04	5	Cotton (secondary)	923 800	18.30
05	6	Pyrethrum	464 900	9.21
06	7	Sisal (secondary)	584 600	11.58
10	8	Tea (primary)	67 500	1.34
12	9	Tea/Coffee	144 000	2.85
13	10	Tea/Sugarcane	55 800	1.11
15	11	Tea/Pyretrum	53 300	1.06
20	12	Coffee (primary)	50 800	1.01
23	13	Coffee/Sugarcane	103 600	2.05
30	14	Sugarcane (primary)	69 200	1.37
34	15	Sugarcane/Cotton	289 400	5.73
40	17	Cotton (primary)	341 000	6.76
60	18	Sisal (primary)	170 900	3.39
70	19	Pineapple (primary)	16 000	0.32
Total extent			5 047 600	100.00

TABLE 3.20
Extents of forest zones

Forest symbol	Forest code	Forest description	Extent (ha)	Percentage area
F1	2	Registered Forest	1 522 000	77.09
F2	3	Unregistered Forest	60 000	3.04
F3	4	Proposed Forest	392 300	19.87
Total extent			1 974 300	100.00

TABLE 3.21
Extents of parkland areas

Parkland symbol	Parkland code	Parkland description	Extent (ha)	Percentage area
P1	2	National park	2 820 300	67.18
P2	3	Game reserve	890 500	21.21
P3	4	National reserve	487 300	11.61
Total extent			4 198 100	100.00

Tables 3.19, 3.20, 3,21, 3.22 and 3.23 present descriptions, map symbols, computer coding and extents of map units respectively of cash crop zones, forest zones, parkland areas, irrigation schemes and tse-tse infestation areas. Figures 3.12, 3.13, 3.14, 3.15 and 3.16 present generalized maps of the above information and the details are presented in Technical Annex 3.

TABLE 3.22
Extents of irrigation schemes

Irrigation area			Extent (ha)	Percentage area
symbol	code	Name		
01	2	Turkwell	400	0.98
02	3	Katilu	700	1.72
03	4	Amolem	400	0.98
04	5	Kaputir	500	1.23
05	6	Bunyala	500	1.23
06	7	Ahero I	2 000	4.91
07	8	Ahero II	2 500	6.14
08	9	Marigat	1 200	2.95
09	10	Mwea	13 200	32.43
10	11	Malka daka	400	0.98
11	12	Merti	400	0.98
12	13	Mbalambala	500	1.23
13	14	Garisa	1 000	2.46
14	15	Hola	2 700	6.63
15	17	Garsen	2 400	5.90
16	18	Tavete	600	1.47
17	19	Mandera	100	0.25
19	20	Bura (proposed)	10 900	26.78
16	21	Wemba	300	0.74
Total extent			40 700	100.00

TABLE 3.23
Extent of tse-tse infestation areas

Tse-tse infestation areas symbol	code	Extent (ha)	Percentage total area
T	2	7 561 000	13.15

3.4 Computerized Land Resources Inventory

The computerized land resources inventory for Kenya records total extents of agro-ecological cells. Each cell contains information on the following:

- Sequence number (NUM)
- Province (PRV)
- District (DIST)
- Thermal Zone (TZ)
- Length of Growing Period Zone (LGP)
- Pattern of Length of Growing Period Zone (PAT)
- Soil Mapping Unit (MPU)
- Landform (LNDFM)
- Geology/Parent Material (GEO)
- Soil Unit (SOIL)

- Soil Texture (TXT)
- Soil Phases 1st, 2nd and 3rd (PHASES)
- Cash Crop Zone (CROP)
- Forest Zone (FOR)
- Irrigation Scheme (IRR)
- Tse-tse Infestation Areas (TSE)
- Parkland Area (PARK)
- Extent in hectares (EXTENT).

The land resources inventory of Kenya consists of about 91 000 unique agro-ecological cells. For Kiambo District (code: 1) in Central Province (code: 1), the complete land resources inventory (651 cells) is presented in Technical Annex 1.

The computerized land resources inventory is available on diskettes (ASCII), in the form of a grid based geographic information system (GIS). For details on formats of the land resources inventory and the use of the geographic information system, reference should be made to Technical Annex 7.

FIGURE 3.12
Generalized map of cash crop zones

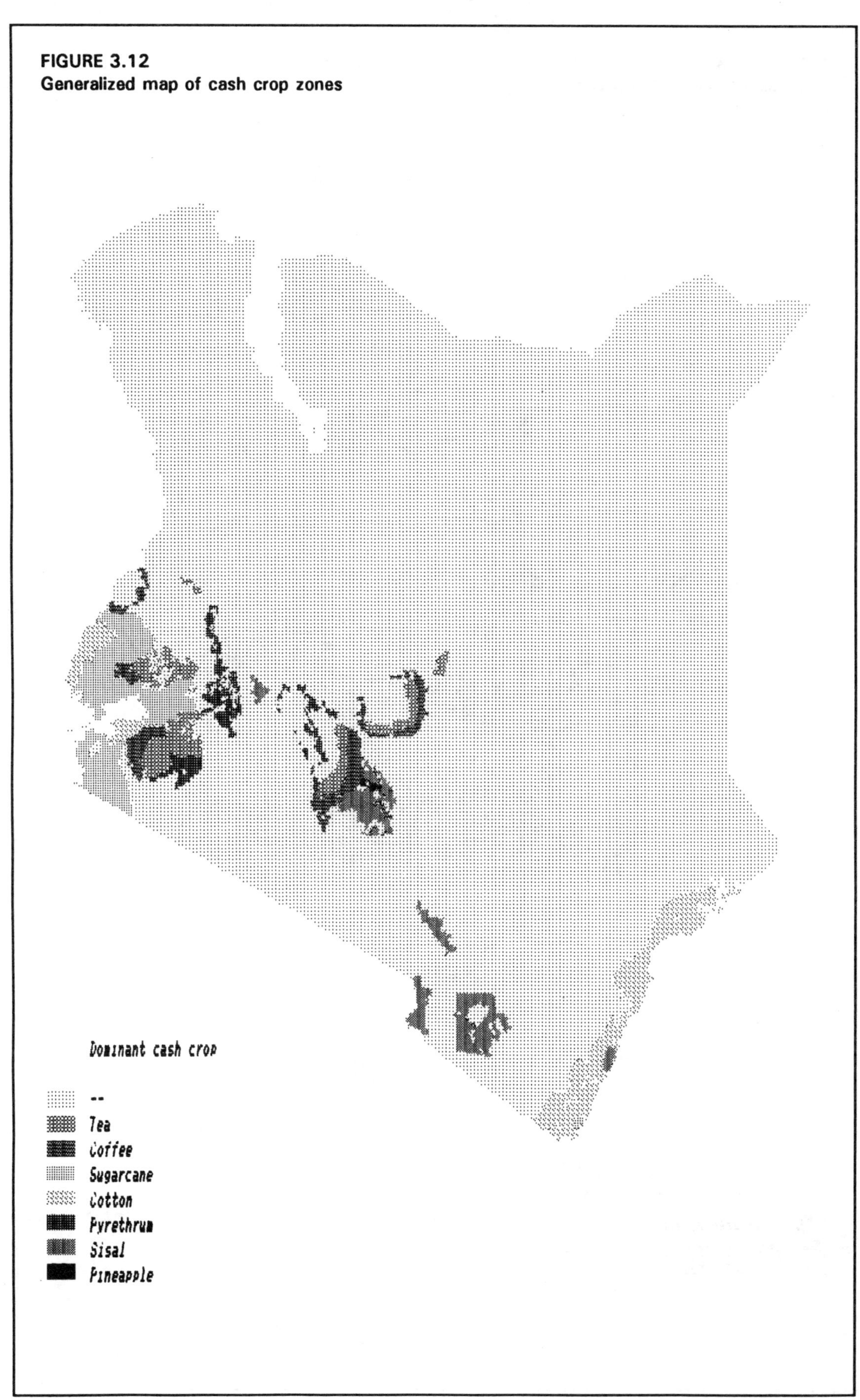

FIGURE 3.13
Generalized map of forest zones

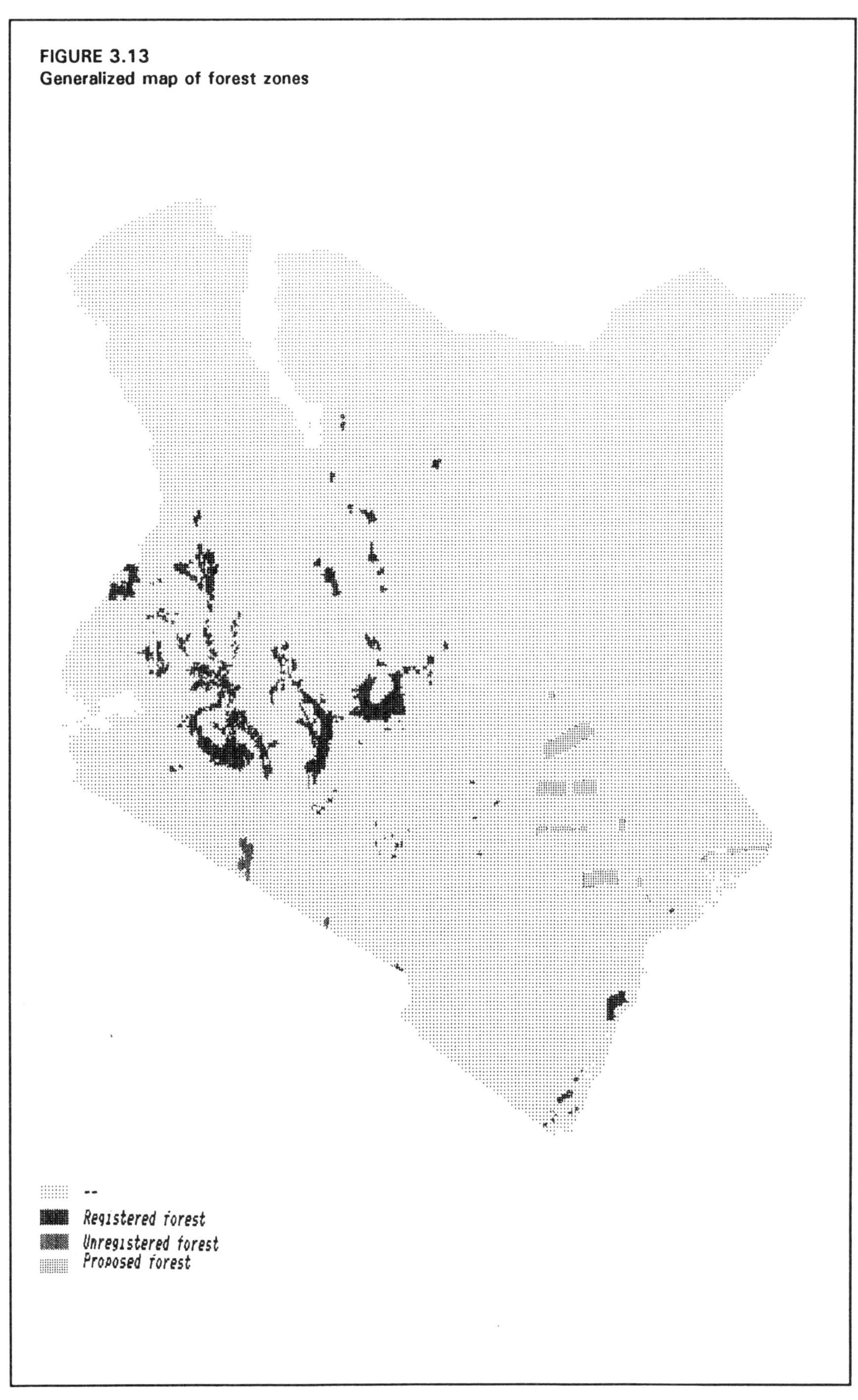

FIGURE 3.14
Generalized map of parkland areas

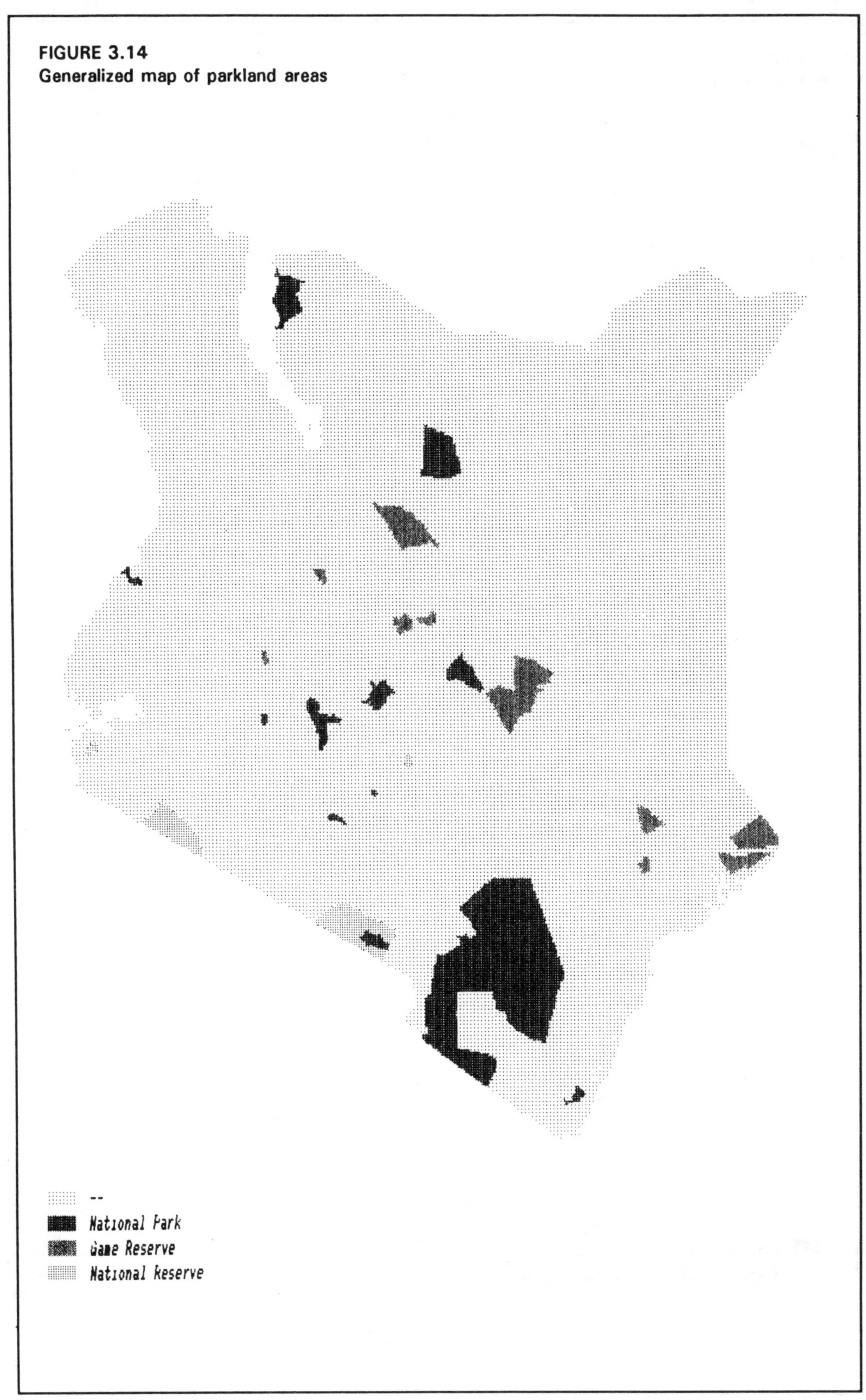

FIGURE 3.15
Generalized map of irrigation schemes

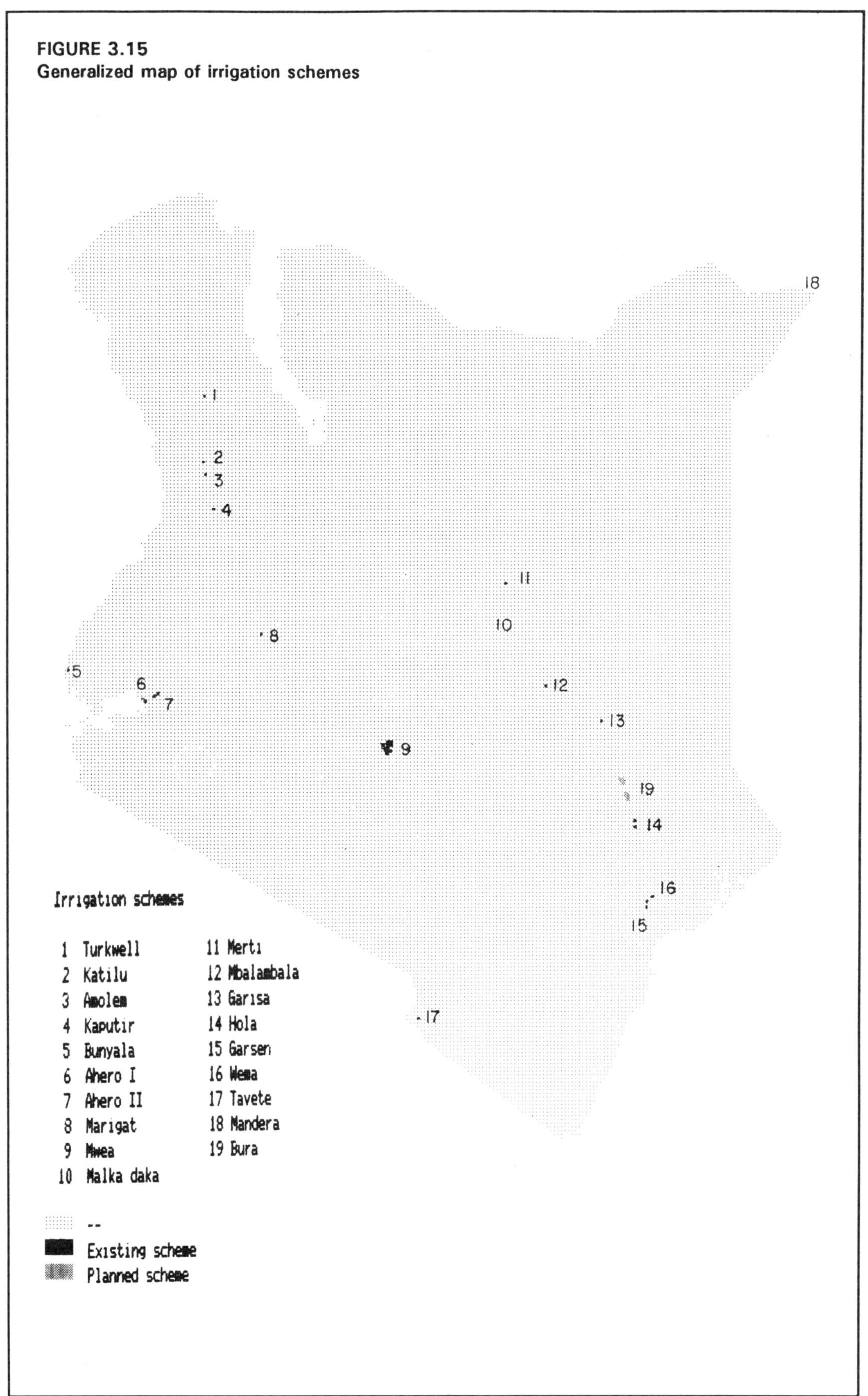

FIGURE 3.16
Generalized map of tse-tse infestation areas

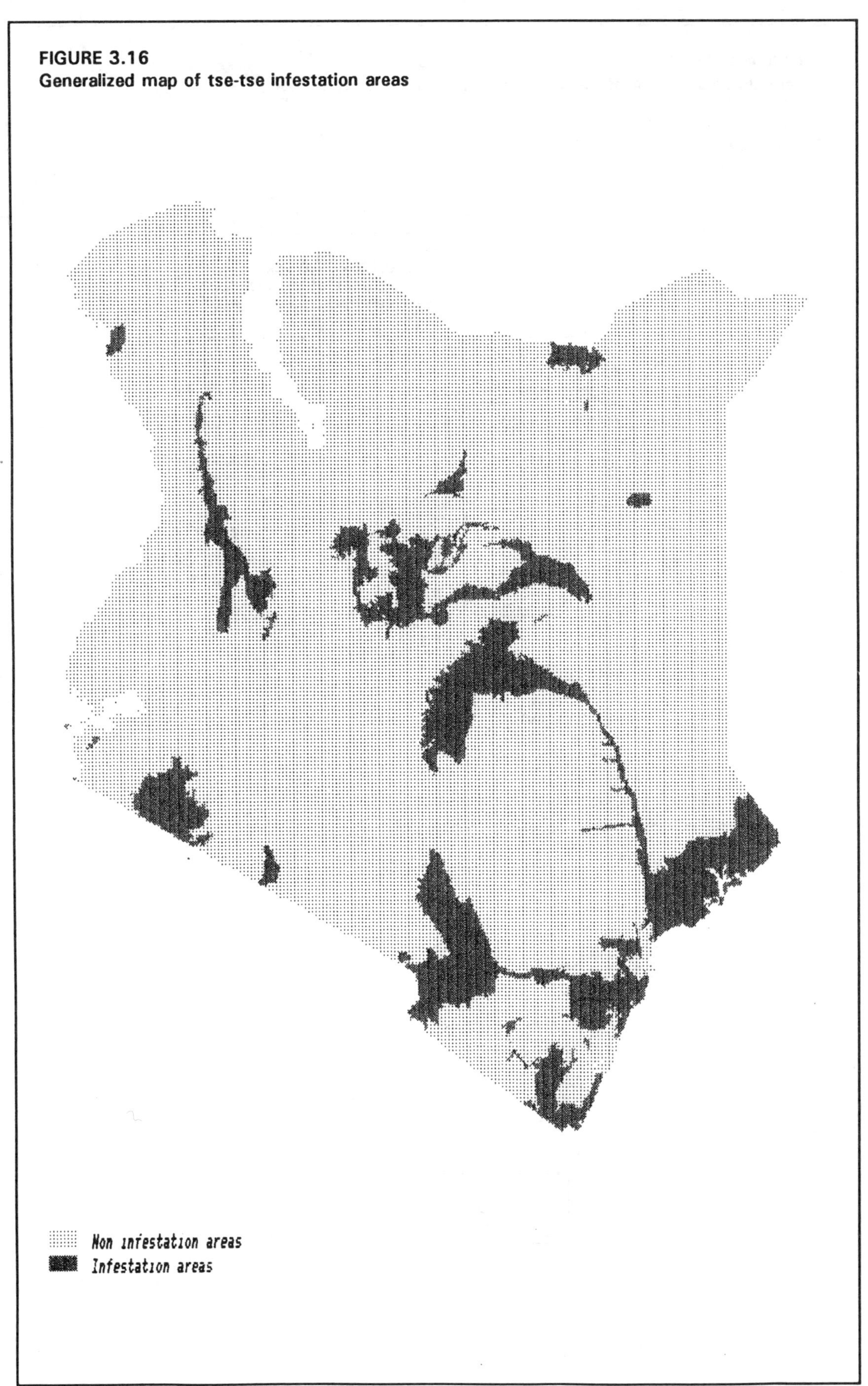

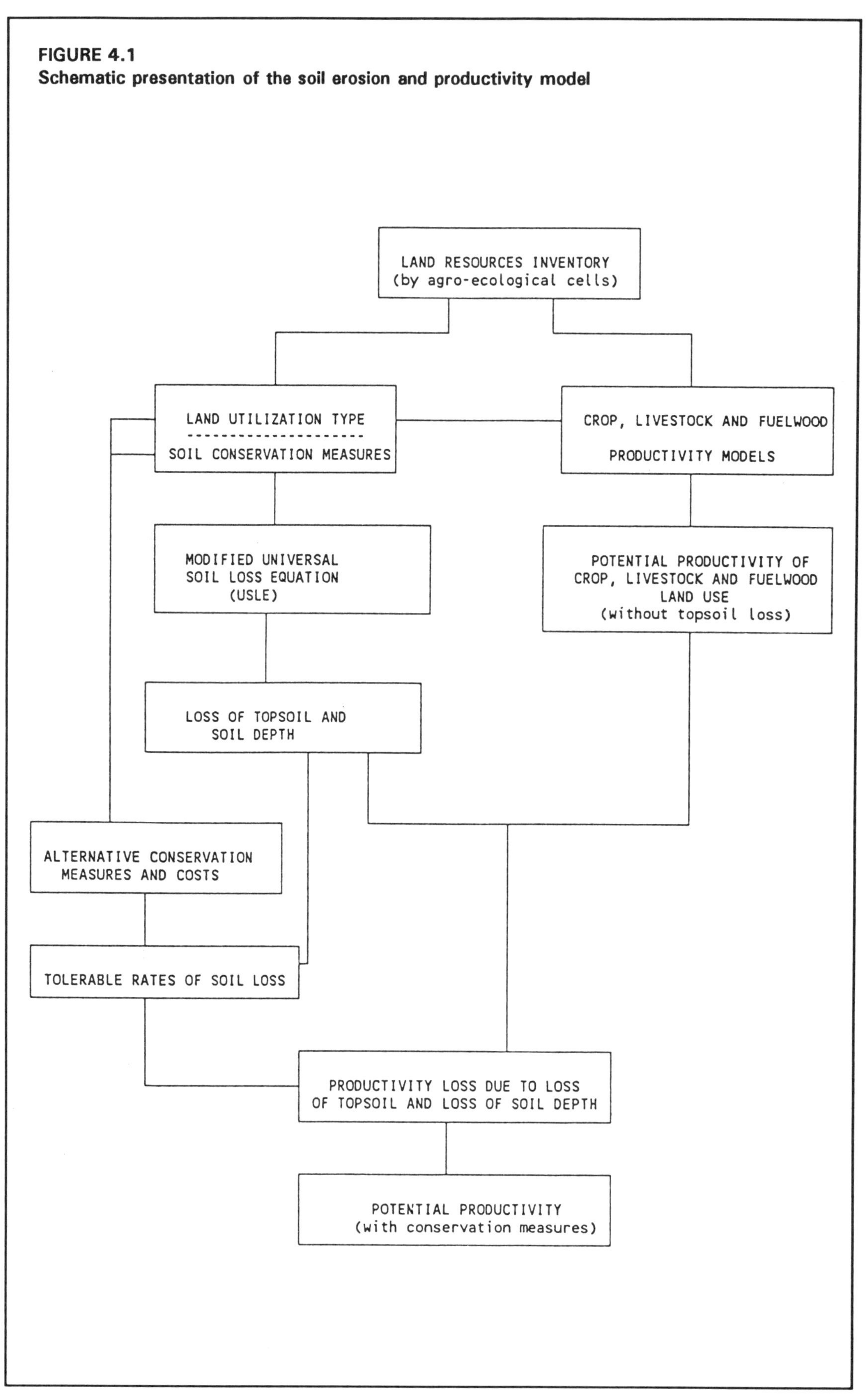

FIGURE 4.1
Schematic presentation of the soil erosion and productivity model

Chapter 4

Soil erosion and productivity

This chapter describes the soil erosion and productivity model, which quantifies implications of alternative land uses in terms of topsoil loss due to erosion and its impact on the productivity of land under different assumed soil conservation measures. The model operates on the climatic and soil resources inventories described in Technical Annex 1. and in Chapter 3. Details of the model are described in Technical Annex 2.

The methodology for the estimation of topsoil loss is essentially based on a modified Universal Soil Loss Equation (USLE): Wischmeier and Smith (1978). The topsoil loss is subsequently converted into productivity loss with or without specific soil conservation measures. The methodology is schematically shown in Figure 4.1. and comprises the following steps:

(i) Identification of land utilization types (LUTs), as defined for crop, livestock and fuelwood productivity models.

(ii) Determination of (USLE) factors for soil erosion: i.e. rainfall erosivity factor, soil erodibility factor, vegetation/crop cover factor, management factor and physical protection factor.

(iii) Application of USLE to quantify (LUT-specific) topsoil loss.

(iv) Establishing the relationships (equations) between loss of yield and loss of topsoil, and classifing soil units according to their susceptibility of yield loss due to loss of topsoil.

(v) Application of equations from (iv) to estimate productivity loss in relation to productivity potentials as quantified by the land use (crop, livestock, fuelwood) productivity models.

(vi) Derivation of productivity estimates, tolerable soil loss and costs for alternative conservation measures.

4.1 Estimation of Soil Loss

The Universal Soil Loss Equation (USLE) equates soil loss per unit area with the erosive power of rain, the amount and velocity of runoff water, the erodibility of the soil, and mitigating factors due to vegetation cover, cultivation methods and soil conservation. It takes the form of an equation where all of these factors are multiplied together:

FIGURE 4.2
Generalized map of potential erosion hazard

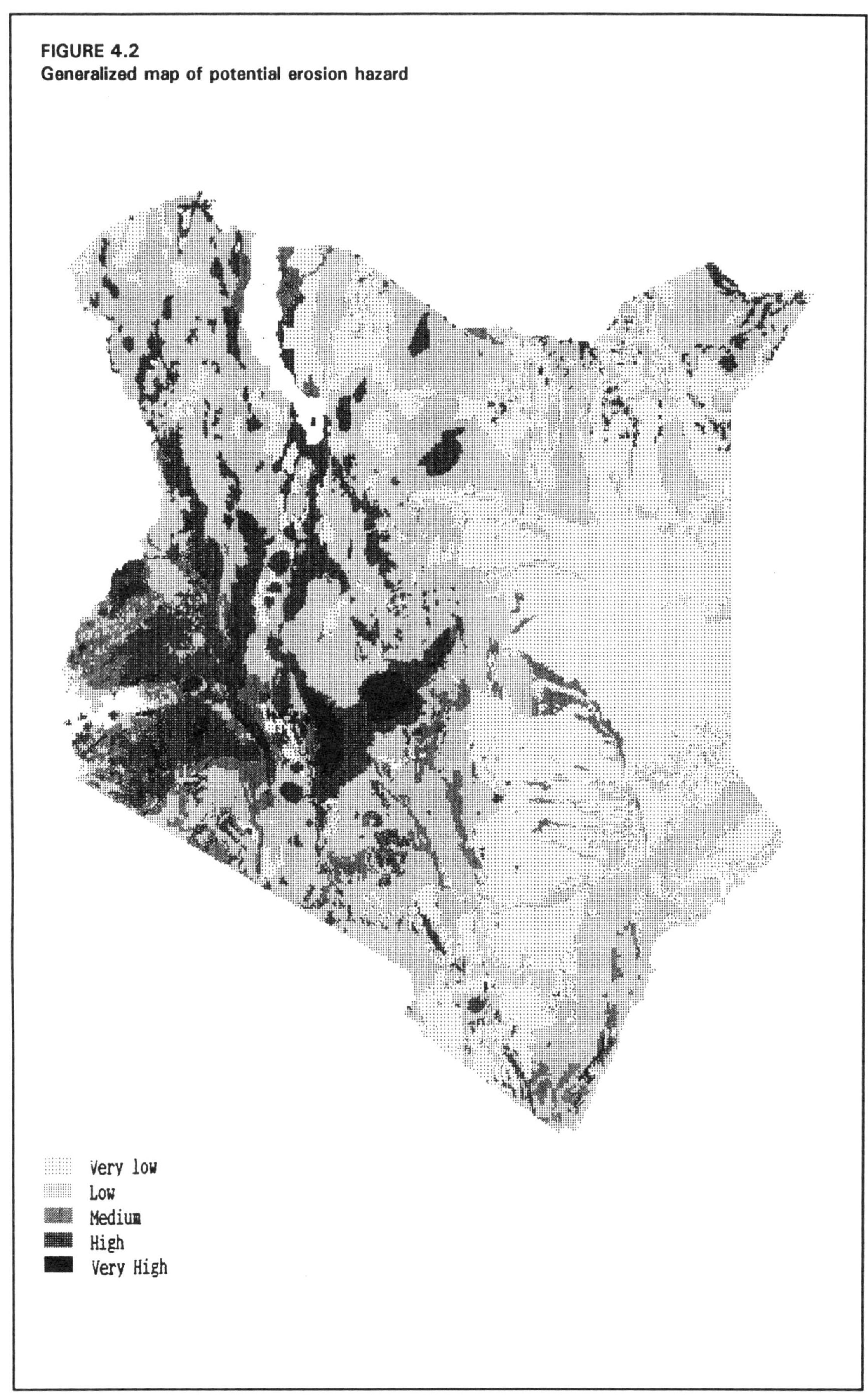

$$A = R \times K \times LS \times C \times P \quad (4.1)$$

where:

A: Annual soil loss in t/ha

R: The rainfall erosion factor to account for the erosive power of rain, related to the amount and intensity of rainfall over the year. It is expressed in units described as erosion index units.

K: The soil erodibility factor to account for the soil loss rate in t/ha per erosion index unit for a given soil as measured on a unit plot which is defined as a plot 22.1 m long on a 9% slope under a continuous bare cultivated fallow. It ranges from less than 0.1 for the least erodible soils to approaching 1.0 in the worst possible case.

LS: A combined factor to account for the length and steepness of the slope. The longer the slope the greater the volume of runoff, the steeper the slope the greater its velocity. LS = 1.0 on a 9% slope, 22.1 m long.

C: A combined factor to account for the effects of vegetation cover and management techniques. These reduce the rate of soil loss, so in the worst case when none are applied, C = 1.0 while in the ideal case when there is no loss, C would be zero.

P: The physical protection factor to account for the effects of soil conservation measures. In this report conservation measures are defined as structures or vegetation barriers spaced at intervals on a slope, as distinct from continuous mulches or improved cultural techniques which come under the management techniques.

The USLE equation has been modified by separating the two elements of the cover and management factor, C as follows:

C*: The vegetation cover factor. This accounts only for the effects of the natural vegetation or crop canopy (including leaf litter and residues accumulating during the life of the crop).

M: The management factor. This accounts for tillage methods, the effects of previous crop residues, previous grass or bush fallows, and applied mulches.

The soil loss equation therefore becomes:

$$A = R \times K \times LS \times (C^* \times M) \times P. \quad (4.2)$$

Equation 4.2 is used in the model to estimate topsoil loss under specified vegetation/crop cover and management conditions for each land utilization type (LUT) as defined in the crop, livestock and fuelwood productivity models (Technical Annexes 4, 5 and 6). These estimates in turn are related to productivity losses and conservation needs (Figure 4.1).

Each of the factors making up the soil loss equation 4.2 is quantified in turn, for a specified land use (LUT) or alternative land uses. Attributes of land utilization types for crops, pasture (livestock) and fuelwood are given in Tables 5.2, 6.2 and 7.2 respectively. The soil loss quantification procedure is presented in Technical Annex 2. Figure 4.2 presents

TABLE 4.1
Slope-cultivation association screen

Land utilization type	Level of inputs Low	Intermediate	High
Dryland crops without soil conservation measures	<30%	<30%	<16%
Dryland crops with soil conservation measures	<30%	<30%	<30%
Wetland crops without soil conservation measures	<5%	<5%	<2%
Wetland crops with soil conservation measures[1]	<30%	<30%	<30%
Coffee, tea, fuelwood and pasture with and without soil conservation measures	<45%	<30%	<45%

1 For wetland crops, terracing is required.

a generalized map of potential erosion hazard. This map combines of the erosivity factor (R), the erodibility factor (K) and the slope factor (LS).

4.2 Soil Erosion and Loss of Productivity

The effect of soil erosion can be measured in different ways according to the kind of damage suffered. In the model (Figure 4.1), the estimate is based on short-term losses in crop production due to erosion of fertile topsoil, and long-term losses in land productivity due to truncation of the soil profile and consequent reduction of available water. No account is taken at this stage in the model development of possible damage to lowlands by flooding and silt deposition, or of the possible benefit from the deposition of fertile silt on alluvial plains, or increase in workability constraints due to changes in terrain characteristics.

In the model, permissible slopes for various land uses under different levels of inputs circumstances have been defined as model variables, and these are given in Table 4.1. The critical slope values in the slope-land use association screen define the upper slope limits to cultivation, and they may be modified as appropriate.

Further, the model takes into account the loss in crop production by soil erosion through:

(a) the removal of topsoil which, in many soils, is the source of most or all the nutrient fertility; and

(b) reducing the overall depth of the soil profile so that eventually the soil water holding capacity and foothold capacity are reduced to a point where it limits yields.

An acceptable rate of soil erosion is considered to be one that over a specified number of years (e.g. 25, 50 or 100):

(a) does not result in a crop yield reduction of more than a specified amount due to loss of topsoil; and

(b) does not result in more than a specified proportion of land being downgraded to a lower class of agricultural suitability due to soil depth reduction.

These two criteria are not interdependent, so that acceptable rate of soil loss is taken as the lower of the two alternatives. The model therefore provides a framework for assessing tolerable soil loss, based on its likely impact on crop yields and the future availability of cultivable land.

The soil erosion and productivity model (Figure 4.1) is linked to crop, livestock and fuelwood productivity models which provide the assessments of land suitabilities and the associated yield potentials for the estimation of tolerable soil loss.

4.2.1 Effect of Topsoil Loss on Productivity

Soils differ in their susceptibility to loss of productivity as the topsoil is eroded. The differences are related to the depth of the topsoil and the amount of nutrient fertility or presence of unfavourable conditions in the subsoil.

Loss of productivity due to topsoil loss can be largely compensated by the use of manure and fertilizer, and low rates of soil erosion are compensated to some extent by the formation of new topsoil. The rate of topsoil formation can vary from < 0.25 mm/year in dry and cold environments to > 1.5 mm/year in humid and warm environments (Hammer 1981; Hudson 1981). Topsoil formation at the rate of 1 mm/year is equivalent to an annual addition of 12 t/ha. Therefore, the rate of topsoil formation has been considered as a factor in the model in assessing loss of productivity and tolerable soil losses. Regeneration capacities of soils used in the model in calculating net loss of topsoil are given in Table 4.2 by moisture and thermal regimes.

Based on experimental evidence (Stallings 1957; Barr 1957; Lal 1976a, 1976b, 1976c; Higgins and Kassam 1981) and analytical data from Kenya Soil Survey (KSS 1975, 1976, 1982b), soil units of the Exploratory Soil Map of Kenya have been classified according to their susceptibility to productivity loss with loss of topsoil, and on the presence of other unfavourable subsoil conditions (Table 4.3). These rankings of susceptibility of the soils are related to actual yield losses, by inputs level, through a set of linear equations given in Table 4.4. The reduced impact of topsoil loss under intermediate and high levels of inputs is due to the compensating effect of fertilizers at their normal rates of use. It is assumed that the benefit of fertilizers is less on the more susceptible soils because of their more unfavourable subsoil conditions.

The tolerable loss rate, for a given soil unit and specified amount and time scale of yield reduction, is calculated in the model as follows:

$$TL = \{(Ra/Rm \times 100 \times B \times Dt) + 3T\} / T \qquad (4.3)$$

where:
- TL = tolerable loss rate (t ha^{-1} $year^{-1}$)
- Ra = acceptable yield reduction (%)
- Rm = yield reduction (%) at the given inputs level when the effective topsoil is all lost
- B = bulk density of the soil (g/cm^3)
- Dt = depth of effective topsoil (cm)
- T = time (years) over which yield reduction is acceptable.

TABLE 4.2
Regeneration capacity of topsoil (mm/year) by length of growing period (LGP) and thermal zone

LGP	Thermal zone								
(days)	T1	T2	T3	T4	T5	T6	T7	T8	T9
< 75	0.5	0.5	0.5	0.5	0.5	0.25	0.25	0.25	0.25
75 - 119	1.0	1.0	1.0	0.5	0.5	0.25	0.25	0.25	0.25
180 - 269	1.5	1.5	1.5	0.75	0.75	0.5	0.6	0.5	0.5
> 270	2.0	2.0	2.0	1.0	1.0	0.5	0.5	0.5	0.5

Derived from Hammer (1981).

TABLE 4.3
Ranking of soils (Kenya Soil Survey) according to their susceptibility to productivity loss per unit of topsoil

Most susceptible	Intermediate susceptible	Least susceptible
Acrisols, except Humic Arenosol	Arenosols	Chernozems
Ferralic Cambisols	Cambisols, except Ferralic Cambisols	Fluvisols
Ferralsols, except humic Ferralsols	Gleysols	Histosols
Ironstone soils	Greyzems	Humic
Lithosols	Humic Acrisols	Andosols
Planosols	Humic Ferralsols	Mollic
Rendzinas	Kastanozems	Andosols
Solonchaks	Luvisols	Vertisols
Solonetz	Nitisols	
	Phaeozems	
	Regosols	
	Vitric Andosols	
	Xerosols	
	Yermosols	

TABLE 4.4
Relationships between topsoil loss and yield loss

Soil susceptibility ranking	Levels of inputs	Equation
Least susceptible	Low	Y = 1.0 X
	Intermediate	Y = 0.6 X
	High	Y = 0.2 X
Intermediate susceptible	Low	Y = 2.0 X
	Intermediate	Y = 1.2 X
	High	Y = 0.4 X
Most susceptible	Low	Y = 7.0 X
	Intermediate	Y = 5.0 X
	High	Y = 3.0 X

Y = productivity loss in percent: X = topsoil loss in cm.

4.2.2 *Effect of Soil Depth Reduction on Productivity*

The rate of soil formation by rock weathering is extremely slow, up to 0.025 mm/year on volcanic rocks in humid areas, and < 0.01 mm/year on basement complex rocks in semi-arid areas (Dunne, Dietrich and Brunego 1978). At the highest rate quoted by Dunne *et al.* (1978), it would take 4,000 years to produce 10 cm of soil. Therefore, the rate at which the soil profile is deepened by rock weathering has not been considered as a factor in the model in assessing tolerable soil losses.

The estimation of the effect of soil depth reduction is based on the assumption that there is no significant loss of productivity until the soil becomes so shallow that shortage of moisture becomes a limiting factor. The critical depth varies according to crop and the climate. Once this critical depth is reached, productivity loss is linear until the soil becomes to shallow to produce any crop at all (Wiggins and Palma 1980). The critical points can be equated with land suitability class limits as follows (where depth is the limiting factor):

VS/S: Soil water becomes limiting and there is at least 20% decrease in yield potential
S/MS: Soil water becomes limiting and there is at least 40% decrease in yield potential
MS/mS: Soil water becomes limiting and there is at least 60% decrease in yield potential
mS/N: Soil water becomes limiting and there is at least 80% decrease in yield potential.

The suitability classes VS (very suitable), S (suitable), MS (moderately suitable), mS (marginally suitable) and N (not suitable) correspond to yield levels of > 80%, 60-80%, 40-60%, 20-40% and < 20% of maximum attainable yield respectively.

If erosion takes place uniformly on soils of varying depth, the end result will be that some soils that had been marginally deep enough will become non-productive while others will become marginal. If the range of soil depths is known, the tolerable amount of soil loss can be gauged in terms of the amount of land that can be permitted to be lost to production.

In order to calculate tolerable soil losses, soil depth reduction is measured in terms of the proportion of the soils in a specified area that have become shallower than a given depth, as a result of erosion. The soils of the mapping units of the Exploratory Soil Map of Kenya (KSS 1982a) are assigned to 5 depth classes: shallow, < 50 cm; moderately deep, 50-80 cm; deep, 80-120 cm; very deep, 120-180 cm; extremely deep, >180 cm.

The rate of soil loss is related to the proportion of land whose soil has become shallower than a specified depth, by the following equations:

(a) Proportion (P, percent) of land downgraded to at least the next depth class:

$$P = (SL \times T) / (B \times Dr) \qquad (4.4)$$

where:
SL = soil loss (t ha^{-1} $year^{-1}$)
T = time (years)
B = bulk density of the soil (g/cm^3)
Dr = depth range of the soil class (cm).

(b) Proportion (P, percent) of land downgraded by more than one depth class:

TABLE 4.5
The proportion of land downgraded from given depth classes to shallower depth classes or to bedrock as a result of soil erosion at different rates over a 100 year period

Soil depth class and change (cm)	Amount of land downgraded (% of class) at erosion rates (t/ha) of:							
	5	10	25	50	75	100	200	400
From shallow (0-50)								
to bedrock (0)	8	17	42	83	100			
From moderately deep (50-80)								
to shallow (0-50)	14	28	70	100				
to bedrock (0)	0	0	0	0	42	100		
From deep (80-120)								
to moderately deep (50-80)	10	21	52	100				
to shallow (0-50)	0	0	0	25	81			
to bedrock (0)				0	0	100		
From very deep (120-180)								
to deep (80-120)	7	14	35	70	100			
to moderately deep (50-80)	0	0	0	3	38	72	100	
to shallow (0-50)				0	0	22	100	
to bedrock (0)						0	78	100
From extremely deep (200-400)								
to very deep (120-180)	2	4	9	19	28	38	76	100
to deep (80-120)	0	0	0	0	1	11	48	100
to moderately deep (50-80)					0	0	30	100
to shallow (0-50)						0	17	92
to bedrock							0	70

TABLE 4.6
Tolerable rates of soil loss (t ha^{-1} $year^{-1}$) to give not more than 10% loss of land from a given depth class and not more than 50% crop yield reduction at low input level over a 100-year period

Soil depth class	Susceptibility to yield loss of topsoil		
	Low	Intermediate	High
Shallow[1]	3.0	3.0	3.0
Moderately deep	3.6	3.6	3.6
Deep	4.8	4.8	4.8
Very deep	7.2	7.2	7.2
Extremely deep	12.0	26.4	26.4

[1] Assuming a minimum depth of 25 cm for crop production.

$$P = \frac{[(SL \times T)/100B]}{Dr} - D2 \times 100 \qquad (4.5)$$

where: D2 = difference (cm) between the lower limit of the depth class and the upper limit of the shallower class to which the land is downgraded

Dr = depth range of the soil class (cm).

Table 4.5 shows the proportions of land downgraded from given depth classes to shallower classes as a result of soil erosion at different rates over a 100 years period. The values in Table 4.5 are based on equations 4.4 and 4.5, and assume that soil depths are evenly distributed over the range in each depth class.

If a tolerable soil loss was set to allow 10% of each depth class to be downgraded by one class over 100 year period, this would give the following soil loss rates for each depth class (assuming 25 cm is the minimum soil depth that would allow crop production):

Shallow (to 25 cm)	- 3	t ha^{-1} $year^{-1}$
Moderately deep	- 3.6	
Deep	- 4.8	
Very deep	- 7.2	
Extremely deep	- 26.8	

4.2.3 Assessment of Tolerable Soil Loss on a Combined Basis of Topsoil Loss and Soil Depth Reduction

Criteria for estimating soil loss tolerance are set according to the amount of yield loss that can be tolerated, or the proportion of the land that can be permitted to become shallower than a specified depth, over a specified time. The two basis for soil loss estimation do not interact, so when used in combination the tolerable soil loss would be the lower of the estimates.

An example of the soil losses that would give either a 50% yield reduction or soil depth reduction resulting in downgrading of 10% of each depth class, over a period of 100 years, is given in Table 4.6.

4.3 Soil Conservation Measures

Three kinds of benifits can be obtained from soil conservation on cultivated land:

(a) Long-term reduction or halting of decline in agricultural production or availability of good quality land.

(b) Immediate or gradual increase in agricultural production.

(c) Non-agricultural benefits such as improved dry season flow of rivers, reduced flooding and siltation of reservoirs, and reduced damage to infrastructure and farm land on lower slopes.

The soil erosion and productivity model essentially quantifies the long-term benefits (i.e. reducing or preventing further losses of agricultural land and decline in crop yields) of seven soil conservation measures for alternative uses of land. It is envisaged that the model would be extended in the future to include an estimation of other agricultural and non-agricultural benefits implied in (b) and (c) above.

The seven types of conservation measures considered in the model are: cut-off drains, narrow-based terraces, bench terraces, converse terraces ('fanya juu ' terraces), grass strips, trash-lines, and stone terraces. Of these, narrow-based terraces and grass strips are suitable for large farms, while all measures except narrow-based terraces are suitable for small farms.

Bench terraces can be used on large farms, but the costs of making them wide enough for mechanical cultivation is high. Also, narrow-based terraces are applicable to slopes < 20%, whereas effectiveness of grass strips is reduced in low rainfall areas (LGP <150 days) because of poor establishment. Also, trash-lines are subject to availability of crop residue whereas stone terraces are subject to availability of stones, and are feasible only on stony soils.

In the application of the soil erosion and productivity model (Figure 4.1), potential erosion losses for each desired land use (crop, livestock, fuelwood) is evaluated first on the assumption that no specific soil conservation measures are applied, i.e. protection factor P = 1. The results are compared with what is considered as acceptable rates of soil loss under the three levels of inputs circumstances, and then the required amount of conservation and associated costs are estimated.

4.3.1 Estimation of Conservation Need

The need for soil conservation is estimated from the protection factor (P) required to reduce soil erosion from its average rate on unprotected land to the tolerable rate estimated in Section 4.2.3. The average rate of erosion covers both the cultivated and the uncultivated parts of the crop and fallow period cycle, but the soil conservation measures described are only applied and maintained in the cultivated part of the cycle. If unacceptable rates of erosion are also occurring during the uncultivated part of the cycle, then additional protection will be needed.

The following example shows how conservation need is estimated in the model, for the cultivated part of the crop and fallow period cycle.

Year		Annual soil loss (t/ha)	Total soil loss (t/ha)
1-4	(Rest period)	4	16
5	(Crop 1st year)	12	12
6	(Crop 2nd year)	18	18
7-10	(Crop 3rd - 6th year)	25	100
Total soil loss over 10 years			146
Tolerable rate of soil loss over 10 years		8	80

Soil loss reduction needed is 66 t/ha (i.e. 146-80). The total soil loss over 6 years of the crop cycle is 130 t/ha, which has to be reduced by 66 t/ ha to 64 t/ha. The P factor needed to achieve this is 64/139 = 0.49.

The protective effect of conservation measures varies according to natural conditions - soil, topography, climate - and the intensity of the measure, e.g. the interval between terraces.

The equations to calculate the required spacing for a given measure, when the relevant natural conditions and the required protection - P factor - are known, are given in Mitchell (1986). These equations form the basis of cost calculations in the model for the seven types of conservation measures listed above.

These conservation measures deal with cultivated land, and in the model their benefit is assumed to last only while the land is under cultivation. If excessive erosion is taking place during the uncultivated part of a crop-fallow period cycle, the most effective way to reduce it is by improving the grass cover. The first requirement for this is to reduce or eliminate the grazing pressure of livestock. Tables 4.7 and 4.8 (Technical Annex 2) show the percent grass cover at different times of the year, and assumed rates of regeneration of grass cover after cultivation, in relation to climate and grazing intensity. These can be used to estimate the effect on soil erosion of reducing the intensity of grazing.

Other possible causes of excessive erosion on uncultivated land are poor established grass due to low rainfall, and unfavourable soil conditions. These have not been explicitly incorporated in the model at this stage in its development, but possible measures to overcome them are: pasture improvement with fertilizers; planting of improved pasture species or broadcasting seed; use of lines of cut bushes to slow runoff to protect germinating grass seeds; use of small earth banks to trap water to encourage germination of broadcast seed (Critchly 1984).

4.3.2 Costs of Conservation Measures

The costs of conservation measures are given in terms of man-days of labour, and the proportion of land taken out of agricultural production by the measures. Where fertilizer is used, for example in establishing grass strips, the amount of fertilizer is specified. It is assumed that all materials used are locally available and therefore not explicitly costed.

Table 4.7 presents a generalized comparison of the characteristics, effectiveness and costs of the seven types of conservation measures considered in the model.

The appropriate conservation measure for a given set of circumstances is normally the cheapest that will achieve the required measure of protection. The costs presented in Table 4.7 are based on man-days of work for manual labour. These include initial costs and maintenance costs. Initial costs are mainly based on the horizontal interval between measures, whereas annual maintenance costs are derived as fixed percentage of the initial costs. In order to compare costs directly, the annual maintenance costs over the cultivated part of the 10-year crop and fallow cycle are converted to net present value using an interest rate of 10%.

Most conservation measures involve taking some land out of production. They vary according to the type of measure, and whether the plants used to protect the terrace banks and other structures have any production value.

TABLE 4.7
Economic aspects of soil conservation measures

Type of measure and physical protection factor (P)	Slope (%)	Horizontal interval (m)	Height of risers (m)	Initial cost			Annual maintenance cost (man-day/ha)	Proportion of land taken out of agriculture (%)
				Construction (man-day/ha)	Grass planting (man-day/ha)	Fertilizer (kg/ha)[1]		
Cut-off drains	>16	-	-	27	-	-	3	-
P = 0.25-0.75[2]	<16	-	-	40	-	-	4	-
Narrow-based	5	40	1.0	50	17	5	5	5
terraces	8	20	1.0	100	36	10	10	10
P = 0.1-0.4[2]	16	10	1.6	200	51	15	20	15
	32	5	1.6	400	102	30	40	30
Bench terraces	12	8	1	1000	44	12	104	6
P = 0.05-0.15[2]	12	16	2	2000	44	12	204	6
	16	6	1	900	58	16	96	8
	16	12	2	1800	58	16	186	8
	24	4	1	750	88	24	84	10
	24	8	2	1500	88	24	159	10
	32	2.8	1	630	125	36	76	13
	32	5.5	2	1270	125	36	140	13
	56	1.4	1	400	250	72	65	22
	56	2.8	2	800	250	72	105	22
Converse terraces	5	20	1.0	100	17	5	18	5
P = 0.05-0.15[2]	8	16	1.3	125	22	6	22	6
	16	8	1.3	250	44	12	44	13
	32	5	1.3	400	72	20	72	20
Grass strips	5	40	-	-	9	2	1	2.5
P = 0.35-0.75[2]	8	20	-	-	18	6	3	5
	16	10	-	-	35	10	5	10
	32	5	-	-	70	20	10	20
Trash-lines	5	40	-	1	-	-	1	2.5
P = 0.35-0.75[2]	8	20	-	2	-	-	2	5
	16	10	-	3	-	-	3	10
	32	5	-	5	-	-	5	20
Stone terraces	5	40	0.4	50	-	-	5	1.5
P = 0.35-0.75[2]	8	20	0.4	71	-	-	7	3
	16	10	0.4	125	-	-	13	6
	32	5	0.4	235	-	-	24	12

1 50% sulphate of ammonia and 50% triple superphosphate.
2 Guideline ranges for physical protection factor (P) under good management only.
Sources: Derived from Mitchell (1986); Vlaanderen (1989).

Chapter 5

Crop productivity

This chapter describes the crop productivity model (Technical Annex 4). The model is schematically shown in Figure 5.1, and comprises five parts:

(i) Land suitability assessment and selection of crop options.
(ii) Formulation of cropping pattern options.
(iii) Formulation of crop rotation options.
(iv) Quantification of productivity potentials of crop rotation options.
(v) Interphase with livestock and fuelwood productivity models.

The model operates on the climate and soil resources inventories described in Chapter 3 and Technical Annex 1. The model explicitly formulates options in respect of annual cropping patterns and crop rotations and quantifies their production potentials at three levels of inputs. The model formulates optimum cropping patterns and quantifies their productivities to meet a given food demand, taking into account desired levels of production 'stability ' at the micro level. The crop productivity model is interphased with livestock and fuelwood productivity models (Chapters 6 and 7).

The five parts of the crop productivity model are described in the following sections.

5.1 Land Suitability Assessment and Selection of Crop Options

Land suitability assessments of single crops (Part I, Figure 5.1) is made according to the FAO-AEZ method (FAO 1978-81), and involves:

(i) selection and definition of land utilization types (e.g. crop, cropping type, produce, production system, inputs level);

(ii) matching the thermal zones of the climatic inventory with the temperature requirements of the crops, and where these requirements are met, computation of agronomically attainable crop yields by length and pattern of growing period zones;

(iii) matching the soil requirements of crops with the soil type, texture classes, stoniness, phases and slope classes of the soil inventory, by rating their limitations.

FIGURE 5.1
Schematic representation of crop productivity model

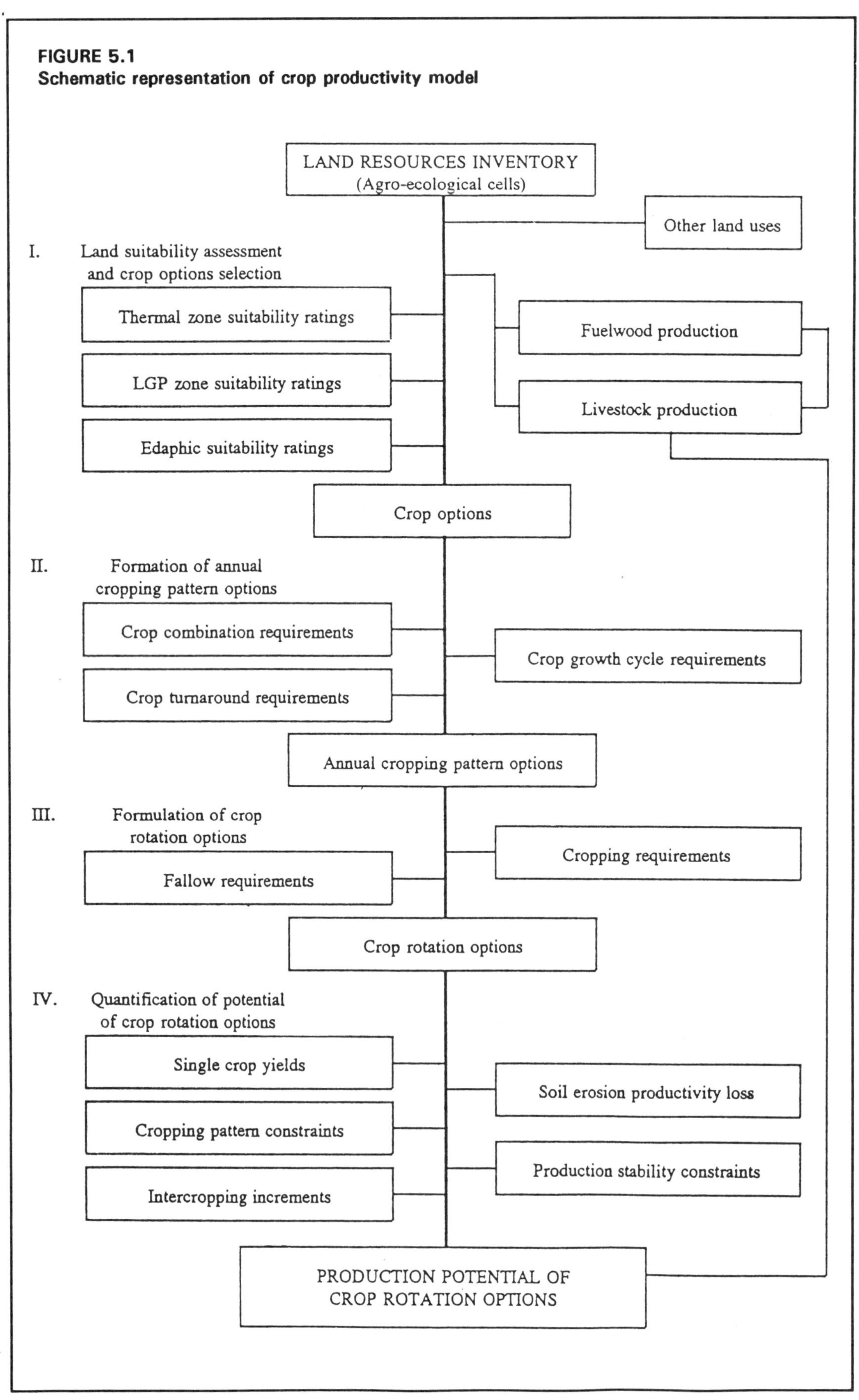

5.1.1 Crops and Land Utilization Types

A total of 25 crop spiecies are considered in the assessment. The full list of crops and crop types is presented in Table 5.1. Coffee, cotton, pineapple, pyrethrum, sisal and tea are considered in the model to take account of the reported production and land area occupied by these crops as quantified by the land use inventory (Technical Annex 3). The remaining 19 crops are differentiated into 58 crop types to account of differences in ecotype adaptation, crop phenology and growth cycles within each crop species.

Each of the 58 crop types are considered at three levels of inputs circumstances (low, intermediate and high). The attributes of the three input level circumstances are listed in Table 5.2, and form the basis of the definition of the land utilization types in the assessment.

The following conditions apply to the crops considered:

(a) cereal and legume crops are grown for dry grain production;

(b) only sorghum varieties with white and yellow grain types are considered;

(c) only maize varieties with white or yellow endosperm types are considered;

(d) barley, oat and wheat cultivars comprise the daylength-neutral types;

(e) groundnut is grown for dry kernel production from either sequentially branched cultivars (the Spanish and Valencia types) or alternately branched cultivars (the Virginia and Castle Cary types);

(f) sugarcane is grown for sugar production using the 'noble ' cane cultivars;

(g) banana is grown for fruit (pulp) production using cultivars from the genome group AAA and AAB;

(h) oil palm is grown for oil production from the fruit mesocarp using the African oil palm stock. It is assumed that the rotation length is 30 years and time to reach first harvest is 6 years.

5.1.2 Climatic Suitability

To enable crops to be matched to climatic conditions, the climatic inventory of Kenya was compiled to permit the interpretation of the climatic resources in terms of their suitability for production of crops. The appropriate climatic adaptability attributes of the crop dictate what parameters need to be taken into account in the compilation of the climatic inventory. The climatic adaptability attributes of crops form the basis of defining the crop climatic requirements, and are outlined in the next section.

5.1.2.1 Crop Climatic Adaptability and Requirements

Crops have climatic requirements for photosynthesis and phenology both of which bear a relationship to yield. The rate of crop photosynthesis and growth are related to the assimilation pathway and its response to temperature and radiation. However, the phenological climatic requirements, which must be met, are not specific to a photosynthetic pathway.

TABLE 5.1
List of crops included in the assessment

Crop	Scientific name	Growth cycle (days)
Barley	*Hordeum vulgare*	90-120
		120-150
		150-180
Maize, lowland	*Zea mays*	70-90
		90-110
		110-130
Maize, highland	*Zea mays*	120-140
		140-180
		180-200
		200-220
		220-280
		280-300
Oat	*Avena sativa*	90-120
		120-150
		150-180
Pearl millet	*Pennisetum americanum*	60-80
		80-100
Rice, dryland	*Oryza sativa*	90-110
		110-130
Rice, wetland	*Oryza sativa*	80-100
		100-120
		120-140
Sorghum, lowland	*Sorghum bicolor*	70-90
		90-110
		110-130
Sorghum, highland	*Sorghum bicolor*	120-140
		140-180
		180-200
		200-220
		220-280
		280-300
Wheat	*Triticum aestivum*	100-130
		130-160
		160-180
Cowpea	*Vigna unguiculata*	80-100
		100-140
		160-190

TABLE 5.1 (Continued)

Crop	Scientific name	Growth cycle (days)
Green gram	*Vigna radiata*	60-80 80-100
Groundnut	*Arachis hypogaea*	80-100 100-140
Phaseolus bean	*Phaseolus* spp.[1]	90-120 120-150 150-180
Pigeon pea	*Cajanus cajan*	130-150 150-170 170-190
Soybean	*Glycin max*	80-100 100-140
Cassava	*Manihot esculenta*	150-300
Sweet potato	*Ipomoea batatas*	115-125 125-145 145-155
White potato	*Solanum tuberosum*	90-110 110-130 130-170
Banana	*Musa spp.*	300-365
Oil palm	*Elais quineensis*	270-365
Sugarcane	*Saccharum officinarum*	210-365
Coffee, arabica	*Coffea arabica*	240-330
Cotton	*Gossypium hirsutum*	160-180
Pineapple	*Ananas comosus*	330-365
Pyrethrum	*Chrysanthemum cinerariaefolium*	210-330
Sisal	*Agave sisalana*	150-270
Tea	*Camelia sinensis*	240-365

1 Includes *P. vulgaris* (Common bean), *P. lunatus* (Lima bean), *P. occineus* (Runner bean) and *P. acutifoleus* (Tepary bean).

TABLE 5.2
Attributes of land utilization types

Attribute	Low inputs	Intermediate inputs	High inputs
Produce and production	Rainfed cultivation of barley, maize, oat , pearl millet, dryland rice, wetland rice, sorghum, wheat, cowpea, green gram, groundnut, *Phaseolus* bean , pigeon pea, soybean, cassava, sweet potato, white potato, banana, oil palm and sugarcane. Sole and multiple cropping of crops only in appropriate cropping patterns and rotations.		
Market orientation	Subsistence production	Subsistence production plus commercial sale of surplus	Commercial production
Capital intensity	Low	Intermediate with credit on accessible terms	High
Labour intensity	High, including uncosted family labour	Medium, including uncosted family labour	Low, family labour costed if used
Power source	Manual labour with hand tools	Manual labour with hand tools and/or animal traction with improved implements; some mechanization	Complete mechanization including harvesting
Technology	Traditional cultivars. No fertilizer or chemical pest, disease and weed control. Fallow periods. Minimum conservation measures	Improved cultivars as available. Appropriate extension packages including some fertilizer application and some chemical pest, disease and weedcontrol. Some fallow periods and some conservation measures	High yielding cultivars including hybrids. Optimum fertilizer application. Chemical pest, disease and weed control. Full conservation measures
Infrastructure	Market accessibility not necessary. Inadequate advisory services	Some market accessibility necessary with access to demonstration plots and services	Market accessibility essential. High level of advisory services and application of research findings
Land holding	Small, fragmented	Small, sometimes fragmented	Large, consolidated
Income level	Low	Moderate	High

Note. No production involving irrigation or other techniques using additional water. No flood control measures.

TABLE 5.3
Average photosynthesis response of individual leaves of four groups of crops to radiation and temperature

Characteristics	Crop stability group[1]			
	I	II	III	IV
Photosynthesis pathway	C3	C3	C4	C4
Rate of photosynthesis at light saturation at optimum temperature (mg $CO_2 dm^{-2} h^{-1}$)	20-30	40-50	> 70	> 70
Optimum temperature (°C) for maximum photosynthesis	15-20	25-30	30-35	20-30
Radiation intensity of maximum photosynthesis (cal $cm^{-2} min^{-1}$)	0.2-0.6	0.3-0.8	> 1.0	> 1.0
Crops included in the Kenya assessment	Barley Oat Wheat Phaseolus bean White potato	Cowpea Green gram Pigeonpea Phaseolus bean Rice Soybean Groundnut Sweet potato Cassava Banana Oil palm	Pearl millet Sorghum Maize Sugarcane	Sorghum Maize

[1] For further information on crop adapatability groups see Tables 3.1 to 3.5 in FAO (1978).

In the FAO Agro-ecological Zones methodology (Kassam, Kowal and Sarraf, 1977), crops are classified into climatic adaptability groups according to their fairly distinct photosynthesis characteristics. Each group comprises crops of 'similar ability' in relation to potential photosynthesis, and the differences between land within groups in the response of photosynthesis to temperature and radiation determine crop-specific biomass productivity when climatic phenological requirements are met.

Crop adaptability groups and their characteristic average photosynthesis response to temperature and radiation are presented in Table 5.3. Barley, oat, wheat, phaseolus bean and white potato have a C3 photosynthesis pathway. They belong to group I and are adapted to operate under cool conditions (<20°C mean daily temperature). Cowpea, green gram, pigeonpea, rice, soybean, groundnut, sweet potato, cassava, banana and oil palm have a C3 photosynthesis pathway. They belong to group II and are adapted to operate under warm conditions (>20°C) with a potential rate of photosynthesis that is greater than in group I crops. Crops in group III (pearl millet, lowland sorghum, lowland maize and sugarcane) have a C4 photosynthesis pathway. They are adapted to operate under warm conditions (>20°C) but with a potential rate of photosynthesis that is greater than in group II crops. Crops in group IV (highland sorghum and highland maize) have a C4 photosynthesis pathway. They are adapted to operate under cool conditions (<20°C) with a potential rate of photosynthesis that is similar to that in group III crops.

TABLE 5.4
Climatic adaptability attributes of crops

Attributes	Barley	Oat	Cowpea	Green gram	Pigeonpea
Species	Hordevum vulgare	Avena sativa	Vigna unguiculata	Vigna radiata	Cajanus Cajan
Photosynthesis pathway	C3	C3	C3	C3	C3
Crop adaptability group	I	I	II	II	II
Days of maturity	90-120[1] 120-150[2] 150-180[3]	90-120[1] 120-150[2] 150-180[3]	80-100[4] 100-140[4]	60- 80[4] 80-100[4]	130-150[4] 150-170[4] 170-190[4]
Harvested part	Seed	Seed	Seed	Seed	Seed
Main product	Grain (C)	Grain (C)	Grain (L)	Grain (L)	Grain (L)
Growth habit	Determinate	Determinate	Indeterminate	Indeterminate	Indeterminate
Life-span Natural	Annual	Annual	Annual	Annual	Short-term perennial
Cultivated	Annual	Annual	Annual	Annual	Annual/Biennial
Yield: Location	TI	TI	LI	LI	LI
Formation period	LT	LT	ME	ME	ME
Thermal zone for consideration	3,4,5,6,7	3,4,5,6,7	1,2,3	1,2,3	1,2,3

C - Cereal
L - Legume
TI - Terminal inflorescence
LI - Lateral inflorescence
LT - Last one third of growth cycle
ME - Middle to end period of growth cycle

Thermal zones:
1 - >25.0 °C
2 - 22.5-25.0
3 - 20.0-22.5
4 - 17.5-20.0
5 - 15.0-17.5
6 - 12.5-15.0
7 - 10.0-12.5

[1] thermal zone 3 and 4
[2] thermal zone 5
[3] thermal zone 6 and 7
[4] thermal zone 1, 2 and 3

The time required to form yield depends on the phenological constraints on the use of time available in the growing period, and the location of yield in the plant (e.g. seed, leaf, stem, root) has an important influence. Temperature has a rate controlling/limiting effect on growth, and it may influence the growth of a specific part and the accumulation of yield if located therein. For example, in wheat, barley and oat, cool night temperatures are required for tillering but the optimum temperatures at the time of flowering and subsequent yield formation are higher. Similarly, optimum temperatures for growth in sugarcane are greater than 20°C but during the ripening period, and because the yield is located in the stem, a lower temperature in the range 10-20°C is required for concentration in the cane of sugar of the right kind. On the other hand, optimum temperatures for growth, development and yield formation in cowpea, green gram and pigeonpea are greater than 20°C and most of the specific temperature requirements are also met when temperatures are optimum for photosynthesis and growth.

The attributes that are helpful in assessing the climatic adaptability of the crops in the matching exercise are given in Table 5.4, for five example crops for adaptability groups I and II. Such information regarding the other crops is given in FAO (1978) and Kassam (1980).

Barley and oat (C3-species, group I) are annuals with a botanically determinate growth habit. Their yield is located in terminal inflorescences in seeds, and the crop yield formation period is the last one-third of their growth cycle. Their climatic adaptability attributes qualify them to be considered for matching in areas with mean daily temperatures less than 22.5 °C and more than 10°C (i.e. thermal zones 3, 4, 5, 6 and 7).

Cowpea (C3-species, group II) is an annual with botanically indeterminate growth habit, offering cultivars that may be morphologically determinate (bunch types) or indeterminate (spreading types). Its yield is located in the lateral inflorescences in seeds, and the crop yield formation period is from the middle to the end of its growth cycle. Its climatic adaptability attributes qualify it to be considered for matching in areas with mean daily temperatures greater than 20°C (i.e. thermal zones 1, 2 and 3).

Green gram (C3-species, group II) is an annual with botanically indeterminate growth habit, offering cultivars that may be morphologically determinate in growth and stature. Its yield is located in the lateral inflorescences in seeds, and the crop yield formation period is from the middle to the end of its growth cycle. Its climatic adaptability attributes qualify it to be considered for matching in areas with mean daily temperatures greater than 20°C (i.e. thermal zones 1, 2 and 3).

Pigeonpea (C3-species, group II) is a short-term perennial with botanically indeterminate but morphologically determinate growth habit. Its yield is located in the lateral inflorescences in seeds, and the crop yield formation period is from the middle to the end of its annual cultivated life-span. Its climatic adaptability attributes qualify it to be considered for matching in areas with mean daily temperatures greater than 20°C (i.e. thermal zones 1, 2 and 3).

5.1.2.2 Thermal Zone Suitability

The association between crop growth cycles and thermal zones in Kenya for the 64 crop types is presented in Table 5.5. In general, growth cycle length (number of days to maturity) of wheat, barley, oat, phaseolus bean and white potato increases by some 5 to 6 days for each 100 m increase in altitude above 1500 m, or for each 0.5 °C decrease in mean temperature from 20.0 °C. In maize and sorghum, there is generally about 20 days extention in maturation for each 100 m increase in altitude above 1500 m or for each 0.5 °C decrease in mean temperature from 20 °C. The 20 days extension in maturity is made up of some 5 to 6 days delay in flowering (silking/anthesis) and some 14 to 15 days extention in the grain filling phase or time taken to reach black layer physiological maturity. For example, 110 and 130 days to maturity correspond respectively to 63 and 69 days to tasseling or heading, 73 and 79 days to silking or anthesis, 110 and 130 days to physiological maturity.

For wheat, barley, oat, phaseolus bean and white potato, mean temperatures of 10 to 12.5 °C or below have been taken to correspond to a risk of frost damage too great for successful cultivation of these crops. Mean temperature range of 10 to 12.5 °C corresponds to 2700 - 3100 m altitude range (Section 3.1.3)

For maize and sorghum, mean temperatures below 15 °C have been considered too low for normal production because of the very severe problems with seed set and maturation. Mean temperatures below 15 °C are reached at altitudes of 2350 m and above (Section 3.1.3).

TABLE 5.5
Crop growth cycle and thermal regime associations

Crop	Growth cycle (days	Thermal regime	
		Range (°C)[1]	Thermal zone
Barley	90-120	17.5-22.5	3,4
	120-150	15.0-17.5	5
	150-180	10.0-15.0	6,7
Maize (lowland)	70-90	> 20.0	1, 2, 3
	90-110	> 20.0	1, 2, 3
	110-130	> 20.0	1, 2, 3
Maize (highland)	120-140	17.5-20.0	4
	140-180	17.5-20.0	4
	180-200	17.5-20.0	4
	200-220	15.0-17.5	5
	220-280	15.0-17.5	5
	280-300	15.0-17.5	5
Oat	90-120	17.5-22.5	3, 4
	120-150	15.0-17.5	5
	150-180	10.0-15.0	6, 7
Pearl millet	60-80	> 20.0	1, 2, 3
	80-100	> 20.0	1, 2, 3
Rice (dryland)	90-110	> 20.0	1, 2, 3
	110-130	> 20.0	1, 2, 3
Rice (wetland)	80-100	> 20.0	1, 2, 3
	100-120	> 20.0	1, 2, 3
	120-140	> 20.0	1, 2, 3
Sorghum (highland)	120-140	17.5-20.0	4
	140-180	17.5-20.0	4
	180-200	17.5-20.0	4
	200-220	15.0-17.5	5
	220-280	15.0-17.5	5
	280-300	15.0-17.5	5
Sorghum (lowland)	70-90	> 20.0	1, 2, 3
	90-110	> 20.0	1, 2, 3
	110-130	> 20.0	1, 2, 3
Wheat	100-130	17.5-22.5	3, 4
	130-160	15.0-17.5	5
	160-190	10.0-15.0	6, 7
Cowpea	80-100	> 20.0	1, 2, 3
	100-140	> 20.0	1, 2, 3
Green gram	60-80	> 20.0	1, 2, 3
	80-100	> 20.0	1, 2, 3
Groundnut	80-100	> 20.0	1, 2, 3
	100-140	> 20.0	1, 2, 3
Phaseolus bean	90-120	17.5-25.0	2, 3, 4
	120-150	15.0-17.5	5
	150-180	12.5-15.0	6

Crop	Growth cycle (days	Thermal regime	
		Range (°C)[1]	Thermal zone
Pigeon pea	130-150	> 20.0	1, 2, 3
	150-170	> 20.0	1, 2, 3
	170-190	> 20.0	1, 2, 3
Soybean	80-100	> 17.5	1, 2, 3, 4
	100-140	> 17.5	1, 2, 3, 4
Cassava	150-330	> 17.5	1, 2, 3, 4
Sweet potato	115-125	> 20.0	1, 2, 3
	125-145	> 20.0	1, 2, 3
	145-155	> 20.0	1, 2, 3
White potato	90-110	15.0-22.5	3, 4, 5
	110-130	12.5-22.5	3, 4, 5, 6
	130-170	10.0-22.5	3, 4, 5, 6, 7
Banana	300-365	> 17.5	1, 2, 3, 4
Oil palm	270-365	> 22.5	1, 2
Sugarcane	210-365	> 17.5	1, 2, 3, 4
Coffee (Arabica)	240-330	15.0-22.5	3, 4, 5
Cotton	160-180	> 20.0	1, 2, 3
Pineapple	330-365	> 17.5	1, 2, 3, 4
Pyrethrum	210-330	10.0-20.0	4, 5, 6, 7
Sisal	150-270	> 17.5	1, 2, 3, 4
Tea	240-365	12.5-22.5	3, 4, 5, 6

1 24-hour mean temperature.

The crop thermal zone suitability ratings for each crop type are presented in Table 5.6. Five suitability classes are employed (i.e. S1, S2, S3, S4 and N), and the ratings apply to all three levels of inputs: where requirements are fully met, the zone is adjudged S1; where requirements are sub-optimal the zone is adjudged S2, S3 or S4; where requirements are not met, the zone is adjudged as N (not suitable).

A rating of S1 indicates that the temperature conditions for growth and yield physiology, and phenological development are optimal and that it is possible to achieve the maximum attainable agronomic yield potential if there are no additional climatic and/or edaphic (including landform) limitations. Ratings of S2, S3 and S4 indicate that temperature conditions for growth and development are sub-optimal and that there would be a suppression of yield potential in the order of 25, 50, and 75 percent respectively. A rating of N indicates that temperatures are not suitable for production of the crop.

TABLE 5.6
Thermal zones suitability ratings

Crop code	Crop	Growth cycle (days)	Thermal zone								
			T1	T2	T3	T4	T5	T6	T7	T8	T9
011	Barley	90-120	N	N	S3	S1	na	na	na	N	N
012		120-150	N	N	na	na	S1	na	na	N	N
013		150-180	N	N	na	na	na	S2	S4	N	N
021	Maize (lowland)	70-90	S1	S1	S1	N	N	N	N	N	N
022		90-110	S1	S1	S1	N	N	N	N	N	N
023		110-130	S1	S1	S1	N	N	N	N	N	N
031	Maize (highland)	120-140	N	N	N	S1	na	na	N	N	N
032		140-180	N	N	N	S1	na	na	N	N	N
033		180-200	N	N	N	S1	na	na	N	N	N
034		200-220	N	N	N	na	na	na	N	N	N
035		220-280	N	N	N	na	S2	na	N	N	N
036		280-300	N	N	N	na	S2	S4	N	N	N
041	Oat	90-120	N	N	S4	S2	na	na	na	N	N
042		120-150	N	N	na	na	S1	na	na	N	N
043		150-180	N	N	na	na	na	na	na	N	N
051	Pearl millet	60-80	S1	S1	S3	N	N	N	N	N	N
052		80-100	S1	S1	S3	N	N	N	N	N	N
061	Rice (dryland)	90-110	S1	S1	S3	N	N	N	N	N	N
062		110-130	S1	S1	S3	N	N	N	N	N	N
071	Rice (wetland)	80-100	S1	S1	S3	N	N	N	N	N	N
072		100-120	S1	S1	S3	N	N	N	N	N	N
073		120-140	S1	S1	S3	N	N	N	N	N	N
081	Sorghum (lowland)	70-90	S1	S1	S1	N	N	N	N	N	N
082		90-110	S1	S1	S1	N	N	N	N	N	N
083		110-130	S1	S1	S1	N	N	N	N	N	N
091	Sorghum (highland)	120-140	N	N	N	S1	na	N	N	N	N
092		140-180	N	N	N	S1	na	N	N	N	N
093		180-200	N	N	N	S1	na	N	N	N	N
094		200-220	N	N	N	na	S3	N	N	N	N
095		220-280	N	N	N	na	S3	N	N	N	N
096		280-300	N	N	N	na	S3	N	N	N	N
111	Wheat	100-130	N	N	S4	S1	na	na	na	N	N
112		130-160	N	N	na	na	S1	na	na	N	N
113		160-190	N	N	na	na	na	S2	S4	N	N
211	Cowpea	80-100	S1	S1	S3	N	N	N	N	N	N
212		100-140	S1	S1	S3	N	N	N	N	N	N
221	Green gram	60-80	S1	S2	S4	N	N	N	N	N	N
222		80-100	S1	S2	S4	N	N	N	N	N	N
231	Groundnut	80-100	S1	S1	S3	N	N	N	N	N	N
232		100-140	S1	S1	S3	N	N	N	N	N	N
241	*Phaseolus* bean	90-120	N	S4	S1	S1	na	na	N	N	N
242		120-150	N	na	na	na	S1	na	N	N	N
243		150-180	N	na	na	na	na	S3	S4	N	N

Crop code	Crop	Growth cycle (days)	Thermal zone								
			T1	T2	T3	T4	T5	T6	T7	T8	T9
251	Pigeon pea	130-150	S1	S1	S3	N	N	N	N	N	N
252		150-170	S1	S1	S3	N	N	N	N	N	N
253		170-190	S1	S1	S3	N	N	N	N	N	N
261	Soybean	80-100	S2	S1	S1	S3	N	N	N	N	N
262		100-140	S2	S1	S1	S3	N	N	N	N	N
311	Cassava	150-130	S1	S1	S2	S4	N	N	N	N	N
321	Sweet potato	115-125	S1	S1	S2	S4	N	N	N	N	N
322		125-145	S1	S1	S2	S4	N	N	N	N	N
323		145-155	S1	S1	S2	S4	N	N	N	N	N
331	White potato	90-110	N	N	S4	S1	S1	na	na	N	N
332		110-130	N	N	S4	S1	S1	S2	na	N	N
333		130-170	N	N	S4	S1	S1	S2	S4	N	N
411	Banana	300-365	S1	S1	S1	S3	N	N	N	N	N
421	Oil palm	270-365	S1	S2	N	N	N	N	N	N	N
431	Sugarcane	210-365	S1	S1	S2	S4	N	N	N	N	N
511	Coffee (Arabica)	240-365	N	N	S3	S1	S4	N	N	N	N
521	Cotton	160-180	S1	S1	S3	N	N	N	N	N	N
531	Pineapple	330-365	S1	S1	S1	S3	N	N	N	N	N
541	Pyrethrum	210-330	N	N	N	S3	S1	S2	S4	N	N
551	Sisal	150-270	S1	S1	S2	S4	N	N	N	N	N
561	Tea	240-265	N	N	S3	S1	S2	S4	N	N	N

5.1.2.3 Length of Growing Period (LGP) Zone Suitability

(a) Individual component LGP suitability:

Potential yields with constraints for individual LGPs were derived according to the method developed by the FAO-AEZ project (Kassam 1977; FAO 1978-81) for all crops except coffee, cotton, pineapple, pyrethrum, sisal and tea. The background details for maize, pearl millet, wetland rice, sorghum, phaseolus bean, soybean, cassava, sweet potato and white potato are given in (FAO 1978); the details for groundnut, dryland rice, sugarcane, banana and oil palm are given in (FAO 1980); the details for barley, oat, cowpea, green gram and pigeonpea are given in Technical Annex 3.

Agronomically attainable yield potentials from the agro-climatic viewpoint (i.e. on suitable soils and terrain) for suitable thermal zones (i.e. thermal zones with S1 rating) are presented in the Appendix in Table A5.1 for high level of inputs, in Table A5.2 for intermediate level of inputs and in Table A5.3 for low level of inputs.

Yields in Tables A5.1, A5.2, and A5.3 refer to single crops which act as building blocks in the formulation of annual cropping patterns and crop rotations, taking into account

TABLE 5.7
Reduction ratings for perennials matched to total length of growing period (LGP, days) L2, L3 and L4

Crop	LGP								
	150-179	180-209	210-239	240-269	270-299	300-329	330-364	365-	365+
Cassava	S2	S2	S1	S1	S1	S1	S1	S1	S1
Banana	N	N	N	N	S1	S1	S1	S1	S1
Oil palm	N	N	N	S1	S1	S1	S1	S1	S1
Sugarcane	N	N	S3	S2	S1	S1	S1	S1	S1

LGP-Patterns and soil-landform constraints. Single crop yields attainable with low inputs are set at 25 percent of those attainable with high inputs. Single crop yields at the intermediate level of inputs are set half-way between the high and low inputs yields.

Yields in Table A5.1, A5.2 and A5.3 apply to normal lengths of growing periods, i.e. growing period with a humid period during which precipitation is greater than full potential evapotranspiration. For intermediate growing periods, i.e. growing period with no humid period, full crop water requirements cannot be met and yield reductions are assumed to be of the order of 50% on all soils except Fluvisols and Gleysols. The percentage of occurrence of intermediate lengths of growing periods in all LGP-Pattern zones is 100% in LGP zone 1-29 days, 65% in LGP zone 30-59 days, 25% in LGP zone 60-89 days, 10% in LGP zone 90-119 days and 5% in LGP zone 120-149 days (Chapter 3).

(b) LGP-Pattern suitability:

All annual crops are matched to individual component length of growing periods, i.e. L1, L2₁, L2₂, L3₁, L3₂, L3₃, L4₁, L4₂, L4₃ and L4₄. The LGP-Pattern evaluation for each annual crop is achieved by taking into account all the constituent component lengths in each LGP-Pattern, thus providing a profile of variability in potential yields over time (e.g. average yield. maximum yield, minimum yield). From such information, it is then possible to set the desired level of yield stability (e.g. in terms of percentage difference between maximum yield and minimum yield or in terms of percentage difference between average yield and minimum yield) in the selection of optimum crops and crop rotations.

Perennial crops (cassava, banana, oil palm, sugarcane) are matched to total lengths of H, L1, L2, L3 and L4, with yield potential downgraded as shown in Table 5.7 for LGP-Patterns comprising L2, L3 and L4. Reduction ratings S1, S2 and S3 in Table 5.7 correspond to zero, 25% and 50% yield reduction respectively due to moisture stress. Rating of N represents unsuitable moisture conditions for crop production.

(c) Cash crops LGP and LGP-Pattern allocation ratings:

For coffee, cotton, pineapple, pyrethrum, sisal and tea allocation ratings by LGP and LGP-Patterns were formulated to enable these crops to be allocated to their suitable climatic zones on suitable soils. The LGP allocation rules at high, intermediate and low levels of inputs are given in Technical Annex 4.

(d) Fluvisols suitability:

Cultivation of Fluvisols is governed by the depth, intensity and duration of flooding which occurs in the low-lying areas of these soils. These flooding attributes are generally controlled, not by the amount of 'on site' rainfall but by external factors such as river flood regime, hydrological features of catchment area and catchment-site relationship.

Additionally, cultivation of these soils is normally confined to post-flood periods, the crops being grown on moisture remaining in the soil profile.

As a result of these factors Fluvisols were rated separately for all crops at high, intermediate and low levels of inputs, and the ratings are presented in Technical Annex 4.

5.1.3 Edaphic Suitability

The edaphic suitability assessment is input-specific and based on:

(i) matching the soil requirements of crop with the soil conditions of the soil units described in the soil inventory (soil unit evaluation); and

(ii) modification of the soil unit evaluation by limitation imposed by texture, stoniness and phase.

As a medium in which roots grow and as a reservoir for water and nutrients on which plants continuously draw during their life cycle, soils are natural resource and valuable economic asset requiring protection, conservation and improvement through good husbandry.

The adequate agricultural silvicultural exploitation of the climatic potential and sustained maintenance of productivity largely depends on soil fertility and management of soil on an ecologically sound basis. Soil fertility is concerned with the ability of the soil to supply nutrients and water to enable crops to maximize the climatic resources of a given location. The fertility of a soil is determined by its both physical and chemical properties whose understanding is essential to the effective utilization of climate and crop resources for optimum production.

In order to assess suitability of soils for crop production, soil requirements of crops must be known. Further, these requirements must be understood within the context of limitations imposed by landform and other features which do not form a part of soil but may have a significant influence on the use that can be made of the soil.

5.1.3.1 Crop Edaphic Requirements and Adaptability

The basic soil requirements of crop plants may be summarized under the following headings, related to internal and external soil properties:

(a) Internal requirements:

- the soil temperature regime, as a function of the heat balance of soils as related to annual or seasonal and/or daily temperature fluctuations;

- the soil moisture regime, as a function of the water balance of soils as related to the soil's capacity to store, retain, transport and release moisture for crop growth, and/or to the soil's permeability and drainage characteristics;

- the soil aeration regime, as a function of the soil air balance as related to its capacity to supply and transport oxygen to the root zone and to remove carbon dioxide;

- the natural soil fertility regime, as related to the soil's capacity to store, retain and release plant nutrients in such kinds and proportions as required by crops during growth;

- the effective soil depth available for root development and foothold of the crop;

- soil texture at the surface and within the whole depth of soil required for normal crop development;

- the absence of soil salinity and of specific toxic substance or ions deleterious to crop growth;

- other specific properties, e.g. soil tilth as required for germination and early growth.

(b) External requirements: in addition to the above internal soil requirements of crops, a number of external soil requirements are of importance, e.g.:

- soil slope, topography and characteristics determined by micro and macrorelief of the soil;

- occurrence of flooding as related to crop susceptibility to flooding during the growing period;

- soil accessibility and trafficability under certain management systems.

From the basic soil requirements of crops, a number of crop response related soil characteristics can be derived. One of these characteristics is, for instance, soil pH. For most crops and cultivars, optimal soil pH is known and can be quantified by a range within which it is not limiting to growth. Outside the optimal range, there is a critical range within which the crop can be grown successfully but with diminished yield. Beyond the critical range, the crop cannot be expected to yield satisfactorily unless special precautionary management measures are taken.

The same holds for other soil requirements of plants related to soil characteristics. Many soil characteristics can be defined in a range that is optimal for a given crop, a range that is critical or marginal, and a range that is unsuitable under present technology.

Table 5.8 presents for five example crops of barley, oat, cowpea, green gram and pigeonpea, optimal and critical ranges of the following soil characteristics: soil slope, soil depth, soil drainage, flooding, texture and clay type, natural fertility (including cation exchange capacity, percent base saturation and organic matter), salinity, pH, free calcium carbonate content and gypsum content. Such information for other crops is given in FAO (1978-81, 1980).

TABLE 5.8
Crop edaphic adaptability inventory

CROP	SLOPE (PERCENT)				DRAINAGE	
	High inputs		Low & Int. inputs		All inputs	
	Optimum	Marginal	Optimum	Marginal	Optimum	Range
Barley	0-8	8-16	0-8	8-24	MW-W	I-SE
Oat	0-8	8-16	0-8	8-24	MW-W	I-SE
Cowpea	0-8	8-16	0-8	8-20	MW-W	I-SE
Green gram	0-8	8-16	0-8	8-20	MW-W	I-SE
Pigeon pea	0-8	8-16	0-8	8-20	MW-W	I-SE

Drainage classes — I = imperfectly drained; MW = moderately well drained; W = well drained; SE = somewhat excessively drained; E = excessively drained.

CROP	FLOODING		TEXTURE			
	All inputs		High inputs		Low & Int. inputs	
	Optimum	Marginal	Optimum	Range	Optimum	Range
Barley	F_0	F_1	L-MCs	SL-MCs	L-SC	SL-KC
Oat	F_0	F_1	L-C	SL-MCs	L-SC	SL-KC
Cowpea	F_0	F_1	SL-SCL	LS-KC	SL-SCL	LS-KC
Green gram	F_0	F_1	L-CL	SL-KC	L-CL	LS-KC
Pigeon pea	F_0	F_1	SL-SCL	LS-KC	SL-SCL	LS-KC

Flooding classes — F_0 = no floods; F_1 = occasional flooding.
Texture classes — MCs = montmorillonitic clay, structured; C = clay (mixed unspecified); KC = kaolinitic clay; SC = sandy clay; SiCL = silty clay loam; CL = clay loam; SCL = sandy clay loam; L = loam; SL = sandy loam; LS = loamy sand.

CROP	DEPTH (cm)		$CaCO_3$ (%)		GYPSUM (%)	
	All inputs		All inputs		All inputs	
	Optimum	Marginal	Optimum	Marginal	Optimum	Marginal
Barley	> 50	25-50	0-30	30-60	0-5	5-20
Oat	> 50	25-50	0-30	30-60	0-5	5-20
Cowpea	> 75	50-75	0-20	20-35	0-3	3-15
Green gram	> 75	50-75	0-25	20-35	0-3	3-15
Pigeon pea	>100	50-100	0-25	20-50	0-3	3-15

CROP	pH		FERTILITY REQUIREMENTS	SALINITY (mmhos/cm)	
	All inputs		All inputs	All inputs	
	Optimum	Range	Range	Optimum	Range
Barley	6.0-7.5	5.2-8.5	moderate	0-8	8-12
Oat	6.0-7.5	5.2-8.2	low/moderate	0-5	5-10
Cowpea	5.2-7.5	5.0-8.2	low/moderate	0-3	3-6
Green gram	5.5-7.5	5.2-8.2	moderate	0-3	3-6
Pigeon pea	5.2-7.5	5.0-8.2	low/moderate	0-3	3-6

TABLE 5.8 (continued)

CROP	ALKALINITY (ESP)	
	All inputs	
	Optimum	Marginal
Barley	0-35	35-50
Oat	0-30	30-45
Cowpea	0-5	8-12
Green gram	0-5	8-12
Pigeon pea	0-5	8-12

TABLE 5.9
Relations between basic soil requirements for crops and soil characteristics

Basic soil requirements	Soil characteristics (soil factors)
Moisture availability[1]	- Effective soil depth - Available soil moisture holding capacity - Drainage
Nutrient availability	- Nutrient availability - Soil reaction
Oxygen availability[2]	- Soil permeability - Drainage
Foothold for roots	- Effective soil depth
Salinity	- Soil salinity
Toxicity	- Soil reaction[3]
Accessibility and Trafficability (workability)	- Topsoil consistency and bearing capacity
Soil tilth for species establishment	- Topsoil consistency and bearing capacity

1 Moisture availability is influenced by climatic factors.
2 Oxygen availability is influenced by inundation and flooding characteristics.
3 Chemical properties of soil parent material may also be involved in some cases.

Many of the soil characteristics listed above and in Table 5.9 are at least partly intrinsically related to the soil. This relationship has guided the definition of optimal and marginal ranges of the various soil characteristics and so simplified the subsequent matching of the different soil units with the inventoried soil requirements of crops.

From the basic soil requirements of crops, a number of responses related to soil characteristics have been derived. The correlation between the basic soil requirements listed above and soil characteristics that can be used as soil factors to rate crop performance is given in Table 5.9.

As explained earlier (Section 3.2.5), the soil units (Table 3.16) have been defined in terms of measurable and observable properties of the soil itself, and specific clusters of such properties are combined into 'diagnostic horizons' and 'diagnostic properties'. They are also used to rate soil suitability.

5.1.3.2 Soil Unit Evaluation

The soil unit evaluation is expressed in terms of suitability ratings based on how far the soil conditions of a soil unit meet the crop requirements under a specified level of inputs. The appraisal is effected in five basic classes for each crop and level of inputs, i.e. very suitable (S1), suitable (S2), moderately suitable (S3), marginally suitable (S4), and not suitable (N).

A rating of S1 indicates that the soil conditions are optimal, and that suppression of potential yields (if any) are assumed to be slight. Ratings of S2, S3 and S4 indicate that soil conditions are sub-optimal for crop production and that potential yields would be suppressed by 25%, 50% and 75% respectively. A rating of N indicates that soil conditions are so severe that the soil unit is not suitable for crop production.

The soil unit ratings for all 25 crops are presented in Technical Annex 4.

5.1.3.3 Texture Evaluation

Soil unit ratings apply as given in Technical Annex 4 if there are no additional limitations imposed by texture. Modifications are required where limitations are imposed by texture.

Soil unit ratings remain unchanged for Arenosols (Q), Albic Arenosols (Qa), Cambic Arenosols (Qc), Ferralic Arenosols (Qf), Calcaro-cambic Arenosols (Qkc), Luvic Arenosols (Ql) and Vitric Andosols (Tv), since coarse texture limitations have been already applied in the soil unit ratings.

Soil unit ratings remain unchanged where textures are medium: fine sandy loam (FSL), sandy loam (SL), loam (L), sandy clay loam (SCL), silt loam (SL), clay'loam (CL), silty clay loam (SICL) and silt (SI); or fine: sandy clay (SC), silty clay (SIC), peaty clay (PC) and clay (C).

In all other cases, i.e. soil units with coarse textures: sand (S), loamy coarse sand (LCS), fine sand (FS), loamy fine sand (LFS), and loamy sand (LS), the soil unit rating is lower by 25% for all crops except for groundnut and white potato.

5.1.3.4 Stoniness and Phase Evaluation

Soil unit ratings apply as given in Technical Annex 4 if there are no additional limitations imposed by stoniness or phase.

Limitations imposed by stoniness and phase are rated using the five basic suitability classes described above. The stoniness and phase modification ratings are presented in Technical Annex 4.

5.1.4 Soil Erosion and Yield Loss

Limitations imposed by slope are taken into account in three steps. Step one defines those slopes which are permissible for cultivation by crop/land use and input level (Table 4.1).

FIGURE 5.2
Schematic presentation of the land suitability assessment programme for crop production

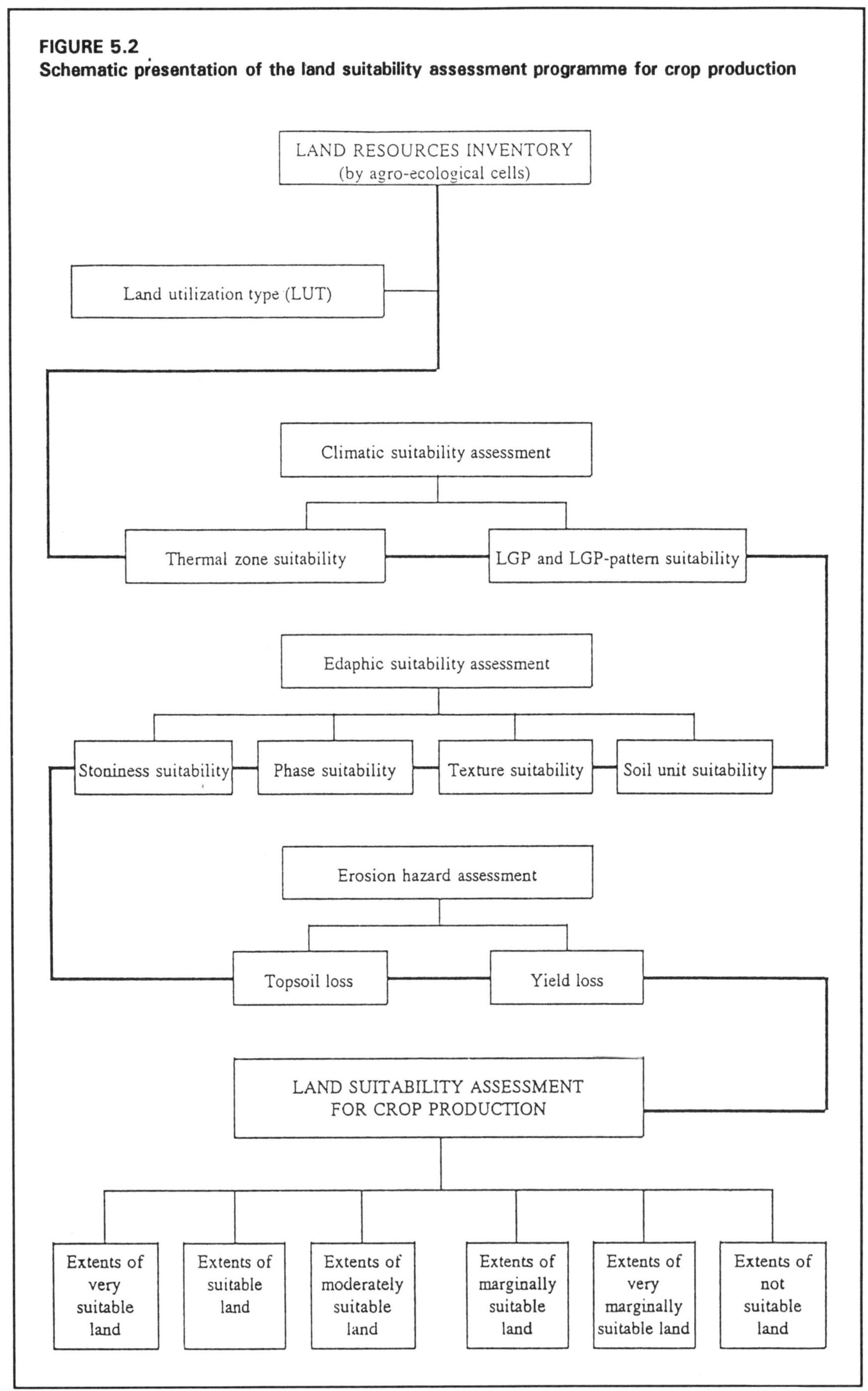

Step two involves the computation of potential topsoil loss, and step three relates topsoil loss to productivity loss according to the method described Chapter 4. Topsoil loss is estimated for each crop and input level using a modified Universal Soil Loss Equation (USLE: Wischmeier and Smith 1978).

The estimated topsoil losses are related to yield losses through a set of equations given in Table 4.4, taking into account the susceptibility of the individual soil units (Table 4.3), level of inputs, and regeneration capacity of topsoil (Table 4.2).

5.1.5 Land Suitability Assessment and Crop Options

The land suitability assessment part of the productivity model (Part I, Figure 5.1) when applied to the land resources inventory (Chapter 3) allows the assessment, by agro-ecological cell, of potential crop performance and consequently crop options to be selected for further processing in Part II of the model. At the same time, land that is reserved for other uses, such as cash crops zones, irrigation schemes, forest zones, reservation and conservation areas, is taken into account as appropriate.

All three assessments: the climatic suitability, the edaphic suitability and the soil erosion hazard, are required to determine the ecological land suitability for crop production of each climate-soil unit of the land resources inventory. In essence the land suitability assessment takes account of all the inventoried attributes of land and compares them with the requirements of the crops, to give an easy to understand picture of the suitability of land for crop production.

The results of the land suitability assessment are presented in five basic suitability classes, each linked to attainable yields for the three levels of inputs considered. For each level of inputs, the land suitability classes are: very suitable (VS) - 80% or more of the maximum attainable yield; suitable (S) - 60% to less than 80% of the maximum attainable yield; moderately suitable (MS) - 40% to less than 60% of the maximum attainable yields; marginally suitable (mS) - 20% to less than 40%; and not suitable (NS) - less than 20%.

Land suitability assessment is achieved by applying the programme illustrated in Figure 5.2. The assessment is carried out separately for each crop and level of inputs.

First, the temperature requirements of the crops with regard to photosynthesis and phenology are compared with the prevailing temperature conditions of each thermal zone. If they do not match, all the growing period zones in that thermal zone are classified as not suitable. If the temperature conditions of a thermal zone partially or fully match with the crop thermal requirements, all growing period zones in that thermal zone are considered for further suitability assessment according to the thermal zone rating.

This further asessment comprises application of length of growing period suitability to the computed areas of the various growing period zones by LGP-Pattern zone. Thus if the thermal zone rating of a particular growing period zone is S1, then potential yield biomass value for the growing period zone is not modified. If the thermal zone rating of the growing period zone is S3, then the potential yield biomass value for the computed extents of the period zone is decreased by 50%. The thermal and moisture suitability assessments are described in Sections 5.1.2 and 5.1.3.

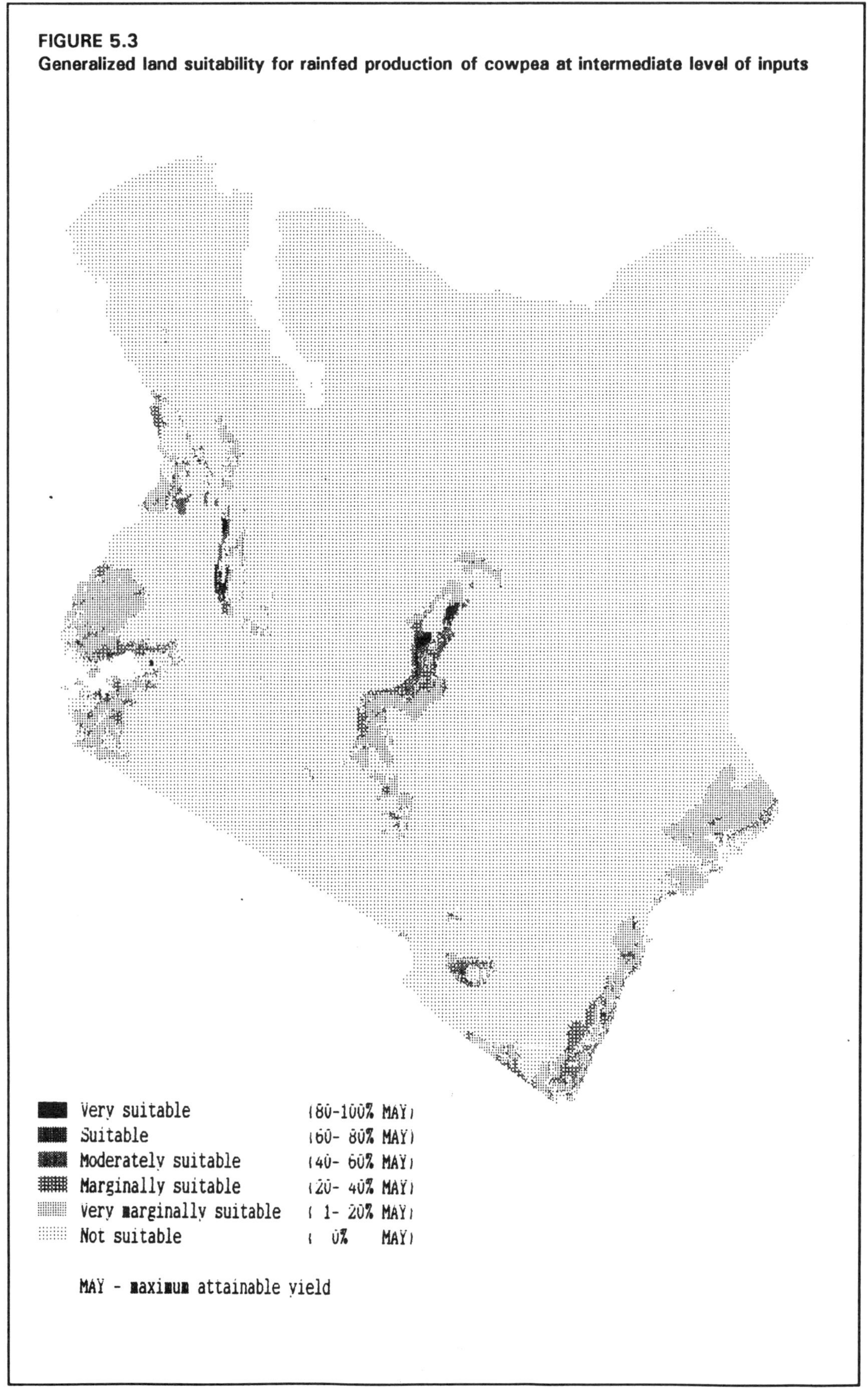

FIGURE 5.3
Generalized land suitability for rainfed production of cowpea at intermediate level of inputs

The length of growing period suitability is applied according to the LGP-Pattern make-up. All annual crops are matched to the individual component length of growing period, i.e. L1, L21, L22, L31, L32, L33, L41, L42, L43 and L44. The LGP-Pattern evaluation for each crop is achieved by taking into account the constituent component lengths of each LGP-Pattern, thus providing a profile of variability in potential yields over time (e.g. average yield, maximum yield, minimum yield). Perennial crops are matched in a similar manner but to total length of growing period, i.e. L1, L2, L3 and L4 and as explained in Section 5.1.2.3.

The next step is an appraisal of the soil units present in each growing period zone. The rating of soil units, for the crops and level of inputs under consideration, is applied to the computed area of the growing period zone occupied by each soil unit. The appraisal, undertaken on the basis of the soil ratings as described in Section 5.1.3, leads to appropriate modifications of the climatic suitability assessment and the attainable yield. Subsequently, the ratings for the different soil textures, phases and stoniness are applied consecutively.

Finally, limitations imposed by slope are taken into account to arrive at the final land suitability appraisal for the crops, for the level of inputs under consideration.

The five classes of land suitabilities are related to attainable yield as a percentage of the maximum attainable under the optimum climatic, edaphic and landform conditions. Consequently the results provide for each land unit an assessment of crop production potentials which can be aggregated for any given area in Kenya.

The generalized results of land suitability assessment at intermediate level of inputs for cowpea are presented in Figure 5.3. and Technical Annex 8. It should be noted that the generalized results presented include a subdivision of the not suitable class (zero to less than 20% of maximum attainable yield) into two classes (1) very marginally suitable (more than zero to less than 20% of maximum attainable yield) and (2) not suitable (zero yield).

For a crop to qualify for selection to the list of crop options for a particular agro-ecological cell, its attainable yield in that cell must be more than the 'threshold' or 'critical' minimum percentage of the maximum attainable yield potential, after taking into account limitations due to climate, soil and erosion hazards.

The threshold minimum yield percentage parameter is a model variable and is set at 20%. Thus, land whose yield potential for any crop is less than 20% of its maximum attainable yield, is regarded as not suitable for production of that crop. Instead, the land would be set aside for subsequent consideration, in Part V of the model, for livestock and fuelwood production (Chapters 6 and 7).

The introduction of a threshold minimum yield potential allows the identification of a 'gross' list of crop options from which a further selection of crops can be made based on additional selection criteria or constraints. In the model, this additional criteria (a model variable) can be set as required, depending on the objective function driving the productivity model.

5.2 Formulation of Cropping Pattern Options

The land suitability assessment part of the crop productivity model (Part I, Figure 5.1) allows the selection of single crops to be made, for each agro-ecological cell, according to their

TABLE 5.10
Definitions of the principle multiple cropping patterns

Multiple cropping	The intensification of cropping in time and space dimensions. Growing two or more crops on the same field in a year.
1. Sequential cropping:	Growing two or more crops in sequence on the same field per year. The succeeding crop is planted after the preceding crop has been harvested. Crop intensification is only in the time dimension. There is no intercrop competition. Farmers manage only one crop at a time in the field.
2. Intercropping	Growing two or more crops simultaneously on the same field. Crop intensification is in both time and space dimensions. There is intercrop competition during all or part of crop growth. Farmers manage more than one crop at a time in the field.

Source: Andrews and Kassam (1976).

TABLE 5.11
Related terminology used in multiple cropping systems

1. Sole Cropping:	One crop variety grown alone in pure stands at normal density. Synonymous with solid planting; opposite of intercropping.
2. Sequential Monoculture:	The repetitive growing of the same sole crop on the same land in a year.
3. Sequential Multiculture:	The repetitive growing of different sole crops on the same land in a year.
4. Cropping Pattern:	The yearly sequence and spatial arrangement of crops or of crops and fallow on a given area.
5. Cropping System:	The cropping patterns used on a farm and their interaction with farm resources, other farm enterprises, and available technology which determine their make-up.
6. Land Equivalent Ratio: (LER)	The ratio of the area needed under sole cropping to one of inter-cropping at the same management level to give an equal amount of yield. LER is the sum of the fractions of the yields of the inter-crops relative to their sole crop yields.

Source: Andrews and Kassam (1976).

yield potentials in the cell. The overall objective of the crop productivity model is to quantify productivity potential of each agro-ecological cell from the whole growing period, or all growing periods per year, and not just single crops from a part of the growing period.

The next step in the model (Part II) is therefore the formulation of cropping pattern options. This is achieved by incorporating the features of multiple cropping using the inputs specific information on land suitabilities and crop options generated through Part I of the model. Once annual cropping pattern options have been formulated, it is then possible to specify cropping pattern constraints and fallow requirements of each cropping pattern.

5.2.1 Multiple Cropping

Multiple cropping is the intensification of cropping in the space and time dimensions, i.e. growing two or more crops on the same land in a year (Table 5.10). The various patterns of multiple cropping reflect essentially two underlying principles - that of growing individual crops in sequence, i.e. sequential cropping; or of growing crops simultaneously in mixtures, i.e. intercropping. Double (and triple, etc.), relay and ratoon cropping work on the former principle, while mixed, row, strip and alley cropping use the latter. There are several other forms of multiple cropping patterns, but these originate through synthesis of the sequential and simultaneous cropping practices. Some of the related terminology used in the multiple cropping systems is given in Table 5.11.

Crops are grown sequentially one after another so that time is used to obtain more production, or crops can be mixed and grown together simultaneously intercropped. With the latter, since the participating crops have different growth requirements, a mixture of crops of similar length to maturity can have higher productivity than a single crop. However, crops commonly used in mixtures usually differ in maturity, so their growth requirements are further separated in time, and competition between them is lower.

The principle of yield increases resulting from a better use of time with crops in sequence is complementary to increases arising from a more efficient use of space with crops in mixture. Theoretically, therefore, maximum cropping should be obtained with sequences of 'high-yielding ' crops in compatible mixtures. In practice, this pattern has evolved in relation to the traditional resources at low and intermediate inputs circumstances where several crops are planted and harvested in mixtures at different times.

The practice of multiple cropping has been reviewed (Andrews and Kassam 1976; Kowal and Kassam 1978; Kassam 1980), and the important rainfed cropping patterns, generalized according to thermal zones and length of growing periods, are given in Table 5.12. Advantages from intercropping are numerous. These include better or fuller use of production resources of water, nutrients, heat, radiation, space and time; better distribution pattern of labour demands, better security of production, better control of pests, diseases and weeds in the absence or sub-optimal use of biocides, better control of soil erosion, and extra yield advantages (i.e. LER > 1.0). Any of these attributes, either singly or in combination, may make intercropping attractive to farmers; and even where there may be no extra yield advantages, intercropping may still be a normal practice because security of production is often a good enough reason for intercropping.

5.2.1.1 Sequential Cropping

Sequential cropping is possible in areas where conditions for crop growth exist beyond the duration of one crop, either due to longer growing period or due to more than one growing period.

In the frost-free areas in Kenya, the restriction to sequential cropping is one of availability of soil moisture. In the areas with a longer growing period, as in the moist sub-humid (growing period 210-270 days) and humid (>270 days) areas, crop growth is possible throughout much of the year. It is in such areas that a strong association with sequential cropping emerges, and sequential crops in both monoculture and multiculture are involved (Table 5.12). However, because of the cool temperatures in the thermal zones T6

TABLE 5.12
Important rainfed cropping patterns generalized according to thermal zones and length of growing period (LGP) zones

LGP	Thermal zone		
(days)	T1, T2, T3	T4, T5	T6, T7
< 120	SC_{as} (I_s) [1]	SC_{as} (I_s)	SC_{as}
120-210	SC_{as} I_s + I_d (S_{mo} + S_{mu})	SC_{as} I_s + I_d	SC_{as} I_s
210-270	SC_{a1} I_s + I_d S_{mo} + S_{mu}	SC_{a1} I_s + I_d (S_{mo} + S_{mu})	SC_{a1} I_s + (I_d)
270-365	SC_{a1} + SC_p I_d + I_s S_{mo} + S_{mu}	SC_{a1} + SC_p I_d + I_s (S_{mo} + S_{mu})	SC_{a1} + SC_p I_s + I_d (S_{mo} + S_{mu})

1 Brackets indicate minor status
SC_{as} Sole cropping of annual short-duration crops
SC_{a1} Sole cropping of annual long-duration crops
SC_p Sole cropping of perennial crops
I_s Intercropping with crops of similar lengths of maturity
I_d Intercropping with crops of different lengths of maturity
S_{mo} Sequential monoculture
S_{mu} Sequential multiculture

and T7, sequential cropping is of minor importance because the annual crops that are adapted to the prevailing conditions are generally slow to reach maturity.

5.2.1.2 Intercropping

In areas with growing periods of less than 120 days, sole cropping of short duration annual crops is dominant in all thermal zones. Some simultaneous cropping is practiced with crops of similar length to maturity, but its status in thermal zones T1, T2, T3, T4 and T5 is a minor one. In the thermal zones T6 and T7 growing conditions only permit a moderate to marginal production from sole cropping of single crops.

In areas with growing periods between 120 and 210 days crop mixtures, including those involving crops of different length to maturity, are common in the thermal zones T1, T2, T3, T4 and T5. Because of the cool temperatures in T6 and T7, crop mixtures involving crops of similar length to maturity are common.

In areas with growing periods greater than 270 days, crop mixtures, especially those involving crops of different lengths of maturity, are common. In such areas, the slow growing and later maturing components generally tend to mature under better end-of-season moisture conditions. In these areas, multiple cropping both on the simultaneous and sequential principle is practised.

5.2.2 Cropping Pattern Options

The cropping pattern options in Part II of the productivity model (Figure 5.1) are formulated by:

(a) Firstly, fitting crop growth cycles into prevailing component lengths of growing periods for each agro-ecological cell. For all annual crops except wetland rice, cropping patterns are made up by fitting growth cycle lengths (full or partial) to total component length of growing period. Where crop growth cycle is curtailed due to inadequate length of growing period, it may still be considered acceptable provided the crop is able to offer a yield.

For wetland rice, growth cycle lengths are matched to the humid period of each component length. For areas with year-round humid lengths of growing periods, the humid period is also year-round. For normal lengths of growing periods, the humid period is approximately 45 days shorter than the total length. For intermediate lengths of growing periods, there is no humid period.

For cassava and the perennial crops of banana, sugarcane, oil palm, the length of crop growth cycle is taken to be equal to the length of the growing period, provided the minimum acceptable growth cycle length can be fitted.

The above matching process is applied, in each agro-ecological cell of the land resources inventory, to all component lengths of growing periods of each LGP-Pattern. This procedure allows the identification of cropping patterns for the complete range of annual cropping periods, including the worst and the best expected, in a given area; and within each year for the different expected lengths of growing periods, including the shortest and the longest.

(b) Secondly, incorporating the 'turn-around' time between crops, within sequential cropping patterns, needed to harvest the first crop, prepare the land and sow the subsequent crop.

The turn-around parameter is a model variable and can be modified as required. For annuals a turn-around period of 10 days has been assumed. For banana and sugarcane a turn-around time of 15 days has been assumed in the year-round growing period zones.

(c) Thirdly, deciding for which levels of inputs and for which crops intercropping is acceptable.

In the model this variable is formulated as follows. Intercropping is considered only at the low and intermediate level circumstance for all crops except wetland rice, sugarcane, banana and oil palm.

5.3 Formulation of Crop Rotation Options

Once annual cropping patterns or crop combinations for each agro-ecological cell in each LGP-Pattern have been formulated, Part III of the crop productivity model (Figure 5.1) formulates crop rotation options. This is done by taking into account crop combination restrictions in space and time of cropping patterns, and fallow requirements of crop combinations that have been selected to participate in the annual cropping patterns.

These two model variables provide for sustainability of production in the longer term. Additionally, they contribute (together with yield limits imposed in the selection of single crops in Part I of the model, and the production stability parameter imposed on the selection of rotations in Part IV) to the overall stability of production system.

5.3.1 Crop Combination Requirements

It is necessary to impose certain crop combination restrictions (as a model variable) for dryland cropping patterns to avoid continuous monocropping. It is also advisable to make a provision for biological nitrogen fixation in the cropping patterns at the low level of inputs circumstance.

At this stage of the model development, the reference crop combination restrictions are that no crop combination should occupy more than two-thirds share of the total cropping area during any year (i.e. total cropping hectare-days available in the agro-ecological cell). The remaining one-third of the annual cropping share of the total hectare-days be occupied by another crop combination according to the following formula.

Where the crop combination (i.e. cropping pattern) occupying the two-thirds of the cropping area is made up of non-legume crops, the remaining one-third of the cropping area should be occupied by a crop combination comprising of legume crops under the low inputs situations. Under intermediate and high inputs circumstances, the latter crop combinations should comprise of non-cereal crops if the former cropping pattern comprise of cereal crops; or non-legume crops if the former are legume crops; or non-tuber crops if the former are tuber crops.

For wetland rice, the above restrictions are not imposed and all of the annual cropping time of an area could be considered for occupation by monocultural cropping pattern, if required. Similarly, the restrictions are not applied to cassava and the perennial crops of banana, sugarcane and oil palm.

The above restrictions for annual and perennial crops may be modified if it is desired that certain proportion of the area of an agro-ecological cell or a group of agro-ecological cells be set aside for specific crops, or where the demand parameter in the objective function imposes a restriction on the types of products that are required.

5.3.2 Fallow Requirements

In their natural state, many soils cannot be continuously cultivated without undergoing degradation. Such degradation is marked by a decrease in crop yields and a deterioration in soil structure, nutrient status and other physical, chemical and biological attributes.

Under traditional farming systems, this deterioration is controlled by alternating some years of cultivation with periods of fallow. The intensity of the necessary fallow is dependent on level of inputs, soil and climate conditions and crops. However the prime reason for incorporating fallows into crop rotations is to enhance sustainability of production through maintenance of soil nutrient fertility.

Nutrient fertility of soils (i.e. the ability of soils to supply nutrients to crops) under traditional subsistance farming (corresponding to LUTs with low inputs) depend mainly on the soil organic matter present in the humus form.

The amount of humus organic matter in soils depends on the relative rates of addition of organic residue and their subsequent breakdown. The relative rates are related to the type, extent and duration of growth of vegetation (natural or crop) and activity of soil organisms, all of which are influenced by soil and climatic conditions.

Maintenance of nutrient fertility of land, cultivated with subsistance low inputs LUTs, is achieved through natural bush or grass fallow as a means of soil fertility regeneration. With intermediate inputs LUTs, providing higher inputs to soils, means of maintaining soil fertility is through fallow which may include for a portion of the time a grass or grass-legume ley or a green manure crop.

Factors affecting changes in soil organic matter are reviewed in Nye and Greenland (1960) and in Kowal and Kassam (1978). They include temperature, rainfall, soil moisture and drainage, soil parent material, and cultivation practices.

The fallow requirements have been derived for the inventoried environmental conditions for four main groups of crops: cereals, legumes, roots and tubers, and banana and sugarcane (Technical Annex 4). The environmental frame used consists of individual soil units, thermal regime, represented by thermal zone T1 (Tmean > 25%), T2 and T3 (Tmean 20-25°C), T4 and T5 (Tmean 15-20°C) and T6, T7 and T8 (Tmean 5-15°C), and moisture regime, represented by length of growing period zones 60-89, 90-119, 120-179, 180-269 and 270-365 days.

Basic values of fallow requirements (F), expressed as percentage of time during the cropping-fallow cycle (i.e. tf/(tc+tf)x100) the land must be put under fallow, for the low inputs LUTs were first calculated. These reference values were then modified depending on the particular crop, and Fertility Capability Classification (Fcc) of the soil (Sanchez, Couto and Buol 1982).

Reference fallow period for LGP > 270 days is 50% greater compared with those for LGP 120-269 days due to additional problems with weeds, pests and diseases, and leaching and erosion. For LGP 90-119 days fallow requirements are greater by 25% due to additional problems with fallow establishment from dry conditions and degradation hazards. For LGP 60-89 days, fallow requirements are greater by 50% due to problems with fallow establishment, degradation hazards and need to conserve moisture.

For moderately warm and moderately cool temperature regimes (T2, T3, T4, T4 and T5 zones) all reference values are decreased by 25% due to lower pest and disease problems and better fallow establishment conditions. For cool temperature regime (T6, T7, and T8 zones), reference values remain unchanged because temperature constraints on the rate of fallow establishment is considered to outweigh any advantage from lower pest and disease infestation.

Fallow requirements for Fluvisols and Gleysols are set lower because of their special moisture and fertility conditions.

Fallow requirements (F) for all suitable soil units are presented in Technical Annex 4 for the low level of inputs situations for cereals, legumes, roots and tubers, and banana and sugarcane.

Fallow requirements at the intermediate level of inputs are taken as one third of those at the low level. At the high level of inputs, fallow requirements are set at 10%.

For wetland rice on Fluvisols, fallow requirements are assumed to be 10% for all the three levels of inputs. For Gleysols, fallow requirements are 40% at the low and intermediate levels, and 10% at the high level.

For long-term perennials (i.e. oil palm, coffee, tea, sisal), fallow requirements are assumed to be nil. For short-term perennials (i.e. pineapple, pyrethrum) and cotton fallow requirements are similar to those for cereal crops.

The fallow factors have been verified against available published data and similar work done earlier by Young and Wright (1980) in the context of FAO's regional assessments.

5.3.3 Crop Rotation Options

In Part III of the productivity model (Figure 5.1), crop rotation options are formulated for each agro-ecological cell for each cropping pattern option generated in Part II of the productivity model. This is accomplished in two steps. Firstly the appropriate crop combination restrictions are applied to rule out risky or undesired crop combinations in space and time, and secondly to incorporate the appropriate fallow requirements for each suitable cropping pattern.

With cropping patterns comprising of more than one crop, average fallow requirements for the crops concerned are applied to define the rotations.

At the same time, Part III of the productivity model defines the extent of fallow land and therefore the portion of biomass that can be used for the livestock production part of the model (Chapter 6).

5.4 Quantification of Productivity Potentials of Crop Rotations

In Part IV of the crop productivity model (Figure 5.1), productivity potentials of land (agro-ecological cells) for each crop rotation option is quantified in three steps.

Step one quantifies the sequential crop yields of each of the crop rotation option. Step two incorporates the intercropping yield increments, and step three applies the production stability constraints (and any other constraints) as criteria for selecting optimum crop rotations and productivities.

Part IV also provides an estimate of potential crop residues, crop by-products and crop primary products that can be made available for livestock production.

5.4.1 Sequential Crop Yields

Sequential cropping is possible in areas where there is either a long continuous growing period or where there are more than one growing period seperated in time due to a marked bimodel (or trimodel) nature of rainfall distribution.

A sequential cropping pattern could be a monoculture or it could be a multiculture. In the former case (e.g. two crops of rice, or white potato), participating crops are of the same adaptability group (Technical Annex 3). In the latter case, the second crop may be different but may belong to the same thermal adaptability group with a similar photosynthesis adaptability response to temperature and radiation (e.g. groundnut followed by cowpea, or pearl millet followed by lowland maize, or wheat followed by white potato) or a different thermal adaptability group (e.g. groundnut followed by lowland maize).

It is therefore an overriding condition that all crops participating in ecologically suitable and desired cropping patterns must first themselves be ecologically suitable. Accordingly, in Part I of the model, a crop type is only permitted to participate in the formulation of a reference crop rotation if its minimum yield (with climate, soil and landform constraints) for the chosen inputs level is more than 20% of its maximum attainable yield. Reference yields, including maximum yields, for situations with no thermal or soil constraints for all crops are given in Tables A5.1, A5.2 and A5.3 in the Appendix for high, intermediate and low level of inputs respectively.

The reference crop yields in Tables A5.1, A5.2 and A5.3 apply (after taking into account climate, soil and erosion constraints) when the crops are considered as single sole crops in the component growing period (i.e. no sequential cropping), or occupy the first position as sole crops in the annual sequential cropping patterns.

Two additional parameters are incorporated in the model (before single crop yields can be applied to quantify sequential crop yields) to take into account:

(i) the increased agro-climatic constraints (e.g. increased pest and diseases, increased workability constraints) on crops when they are positioned second or third in the cropping sequence instead of being first; and

(ii) those situations when the yield formation period of the crop in the cropping sequence cannot be fully accommodated within the time available for cropping, with the consequence that there is a partial yield loss (as opposed to total crop loss).

To allow for the increased agro-climatic constraints for crops in cropping patterns with two or three crops, single sole crop yields of the second and the third crop are downgraded, as shown in Table 5.13 in comparison to the yield as the first crop. Of the 58 crop types considered in the model, cassava, sugarcane, banana and oil palm do not have the possibility of taking up a second crop position in an annual cropping pattern.

Where crops cannot complete there yield formation within the time available, yield reductions are made proportiontely to the decreases in the yield formation periods. The yield formation periods for cereal, legumes and roots and tuber crops are assumed as one-third, one-half and two-thirds respectively of their corresponding total length of normal growth cycle. This assumption is also made in defining the climatic adaptability of crops and in the calculation of net biomass and yields of crops (Technical Annex 3).

TABLE 5.13
Yield reductions (%) of crops when grown as second or third crop in the annual cropping pattern relative to its yield as a first crop

	Crop yield as first crop (% of maximum)			
Crop position	20 - 40	40 - 60	60 - 80	80 - 100
2nd	50	25	25	25
3rd	75	50	25	25

TABLE 5.14
Suggested land equivalent ratios (LER) at different levels of inputs by length of growing period (LGP) and crop yield relative to maximum attainable yield

	Inputs/Relative crop yield									
LGP	Low					Intermediate				
(days)	< 0.2	0.2-0.4	0.4-0.6	0.6-0.8	0.8-1.0	< 0.2	0.2-0.4	0.4-0.6	0.6-0.8	0.8-1.0
< 120	1.0	1.0	1.0	1.0	1.0	1.0	1.0	1.0	1.0	1.0
120-170	1.0	1.1	1.2	1.2	1.3	1.0	1.05	1.1	1.1	1.15
> 270	1.0	1.2	1.3	1.3	1.4	1.0	1.1	1.15	1.15	1.2

5.4.2 Intercropping Increments

The extent of the extra contribution to production per unit area from intercropping has been described in Kassam (1980). In practice, farmers select compatible mixtures with LER values of greater than 1.0 except in situations where intercropping is still advantageous for reasons other than extra yields.

Reference LER values which have been applied in the model are given in Table 5.14. Based on evidence from surveys and experiment, it is assumed that intensifying crop production through intercropping would have its limits. At the high inputs level, the primary disadvantage of intercropping is the difficulty in mechanization, and in effectively conducting some of the cultural operations. This generally restricts the widespread use of intercropping in large farm systems, particularly when under such systems most of the advantages of intercropping no longer apply. It is, therefore, suggested that at high inputs level, there should be no extra yield advantages (LER=1.0) in production over and above that which is already reflected by sole crop yields.

At the low inputs level, the most complex patterns that are also potentially the more productive would eventually require so much labour and other resources that even the small farmer may only use them occasionally on a small part of his land. Further is is most likely that the extra yield advantages claimed under experimental conditions (e.g. LER=1.3-1.5 in LGPs 120-270 days, LER=1.85 in LGPs >270 days) with their accompanying 'high' inputs would decrease by about 50% under field conditions at the low inputs level. However, it is postulated that the extra yield advantages from intercropping would increase with the increase in length of growing period; and that the maximum advantages should be with mixtures where the individual component crops are very suitably adapted to the prevailing climate and soil environment.

It is therefore suggested (as model variables in Table 5.14 that for LGP zones with less than 120 days, there would be no significant extra yield advantages (LER=1.0) from intercropping. The single sole crop yields are considered to adequately reflect the production potential.

For LGP zones with more than 120 days, the following has been applied in the model for all crops except wetland rice, sugarcane, banana and oil palm. For LGP zones 120 to 270 days under low inputs level, there is a 30% extra yield advantage (LER=1.3) from intercropping when attainable yields (from Part I of the model) of the individual participating crops are 80% or more of the maximum attainable yields. This yield advantage is reduced to nil (LER=1.0) for participating crops with attainable yields that are less than 20% of the maximum attainable.

For LGP zones with more than 270 days, there is 40% extra yield advantage (LER=1.4) from intercropping when the attainable yields of the individual participating crops are 80% or more of the maximum attainable yields. This yield advantage is reduced to nil (LER=1.0) for participating crops with attainable yields that are less than 20% of the maximum attainable.

For intermediate inputs level, yield advantages from intercropping are taken as half of those at the low inputs level.

For wetland rice, sugarcane, banana and oil palm, LER of 1.0 has been applied.

5.4.3 Production Stability Constraints

It is necessary to state the level of production stability, e.g. the tolerable difference between minimum (worst year) and maximum (best year) production; the tolerable soil erosion rate, desired from the cropping patterns and rotations that are selected to meet food and other demands specified by the objective function. These constraints are introduced as model variables in the selection of cropping patterns and crop rotations and the quantification of production therefrom.

In the model, the desired level of production stability between minimum and maximum production is set at 75%. This means that the production variations from year-to-year from the selected cropping patterns would not exceed 25%.

5.4.4 Crop Productivity Potential

When the crop productivity model is applied to the land resources inventory, crop productivity potentials of each agro-ecological cell are quantified taking into account the requirements and constraints imposed at the various stages in the model. Results of the crop productivity assessment at district level are presented in Technical Annex 8.

To be able to use the crop productivity potentials for planning, it is necessary to take into account the waste factor and the amount of production which is required as seed and therefore not available to enter the animal and human food chain.

The waste factor, covering post harvest losses during food processing and in the food delivery system, has been taken as 10% and can be varied as required.

The seed factors for crops are given in Technical Annex 4, and are part of the input technology matrix (Bruinsma et al. 1983; Technical Annex 7). They are applied after applying the waste factor, to arrive at the net production available.

The demand for food is expressed in terms of calorie and protein. The calorie and protein conversion factors for food products are given in Technical Annexes 4 and 7.

Once crop productivity potentials are quantified, it is possible to quantify crop residues, crop by-products (groundnut and soybean cake), and crop primary products (grain) which may be or need to be made available for livestock production (Chapter 6).

5.5 Interphase with Fuelwood and Livestock Productivity Models

The last part of the crop productivity model (Figure 5.1) deals with the interphase with fuelwood and livestock productivity models The interphase in essence allows the possibility of considering:

(a) fuelwood production on land assessed as not suitable for crops in Part I of the model;

(b) fodder from fuelwood trees for livestock production;

(c) fallow land, defined in Part III of the model, for livestock production because of the fodder potential of fallows; and

(d) crop residue, crop by-products and crop primary products, quantified in Part IV of the model, for livestock production.

5.5.1 Fuelwood Productivity Model

The fuelwood productivity model is described in Chapter 7. It is basically a land suitability assessment model (similar to Part I of the crop productivity model) in which land potentials for individual fuelwood tree species at three levels of inputs are quantified.

Additionally, any portion of crop land may be considered for fuelwood production depending on how much land is required for crops, livestock and other land uses.

5.5.2 Fodder from Fuelwood Land

Any land which is allocated to fuelwood production with species that offer palatable fodder would have the potential for contributing a portion of the fodder for livestock production.

Fuelwood species which offer palatable fodder to livestock are listed in Chapter 7. The amount of fodder which can be utilized by stock without affecting fuelwood yields would depend on the species and the ecological situation. However, at this stage of the model interphase development, it is assumed that about 10% of the foliage may be utilized by stock without affecting fuelwood yields. This nominal value may be modified as appropriate according to species and environment.

5.5.3 Fodder Potential of Fallow Land

It is assumed that fallow land under the low level of inputs situation would be under bush or natural grass vegetation, whereas under the intermediate and high inputs situations, fallow land would be under sown pastures.

Biomass potentials for sown or natural permanent pastures is assessed in the livestock productivity model described in Chapter 6.

At this stage of the model developement, it is assumed, as a model variable, that the biomass potential from natural grass fallow under low inputs and the sown grass fallow under intermediate and high inputs, is one-third of that quantified for normal permanent or sown pastures. Also, it is assumed that only 50% of the biomass may be consumed by stock.

5.5.4 Crop Residues, Crop By-products and Crop Primary Products

In areas with more than 120 days growing period, crop residues are an important source of fodder particularly for the low and intermediate technology livestock systems. Important residues are the haulms of the groundnut crop, cowpea and other legume crops, and the staves of sorghum, maize and millet and straw from rice, wheat, barley and oat.

Quantities of residues that may be available can be estimated by applying the residue factors and the corresponding utilization coefficients onto crop yields form Part IV of the crop productivity model. Residue factors and utilization coefficients for crops are given in Technical Annex 5.

By-products, defined as edible materials remaining after a crop has been processed, are bran, pollard and germ meal from cereal milling, molasses and bagasses from sugar milling, and cakes (cotton, soybean, groundnut) from oilseeds.

Quantities of crop by-products that may be available can be estimated by applying the by-product factors and the corresponding utilization coefficients onto crop yields. By-product factors and utilization coefficients for crops are given in Technical Annex 5.

Crop primary products apply to grain used for the purpose of feeding to animal either directly in an unprocessed form or in a processed form. Main cereals used in Kenya are maize, sorghum, wheat and barley.

Direct feeding of grain is used in cattle and goat dairy systems, and the intensive livestock industries of poultry and pig production tend to use processed feed. The amount of primary products in the livestock feed requirements for the different livestock production systems are given in Chapter 6.

FIGURE 6.1
Schematic presentation of the livestock productivity model

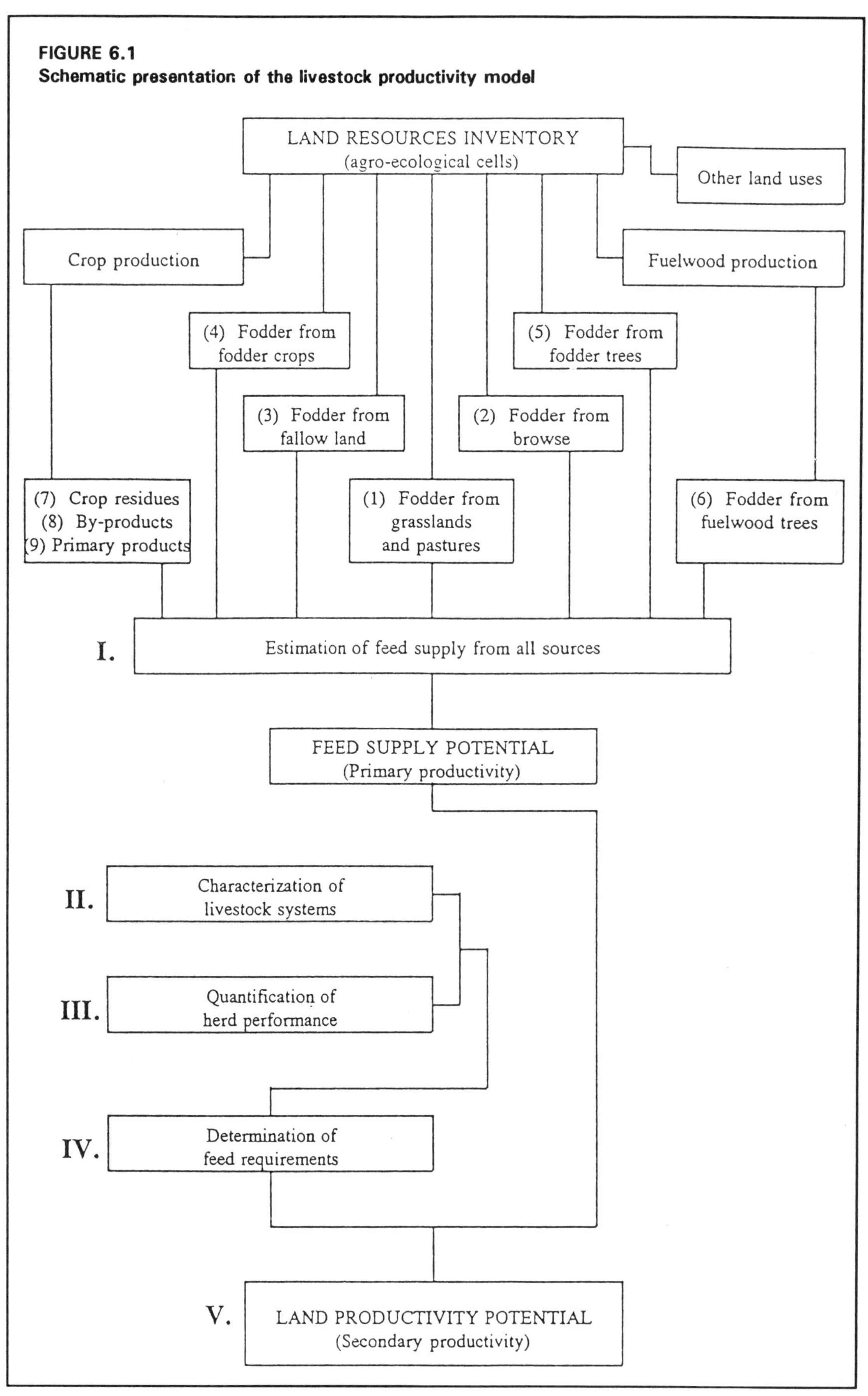

Chapter 6

Livestock productivity

This chapter describes the livestock productivity model (Technical Annex 5). The livestock productivity model is schematically shown in Figure 6.1. It has been conceptualized and applied within the framework of land evaluation guidelines (FAO 1976, 1988a), and follows the FAO Agro-ecological Zones (FAO-AEZ) approach to quantifying land resources and assessing land use potentials (FAO 1978-81; Blair Rains and Kassam 1980).

The livestock productivity model has five parts, namely:

(i) Estimation of feed supply potential (primary productivity).
(ii) Characterization of livestock systems.
(iii) Determination of herd performance.
(iv) Estimation of feed requirements.
(v) Quantification of livestock productivity potential (secondary productivity).

The model operates on the land resources inventory described in Chapter 3. The five parts of the model, are described in the following sections.

6.1 Estimation of Feed Supply

Animals require a continuous and adequate supply of nutritively satisfactory feed. Part I of the livestock productivity model (Figure 6.1) deals with the estimation of feed supply from different sources. A wide variety of plant biomass is eaten by domestic herbivores. The important sources of feed biomass are the grasses, a small number of herbaceous legumes, leaves and fruits of many shrubs and trees, fodder crops, crop residues, crop by-products and primary products (e.g. grain).

6.1.1 Sources of Feed

At any given location, the ecological potential of one or more of the following sources of feed needs to be quantified by agro-ecological cells of the land resources inventory (Figure 6.2).

(i) Grassland or pastures (permanent or long-term, or short-term grass-legume mixtures, natural or sown).

(ii) Browse (natural woody vegetation of shrubs and trees).

FIGURE 6.2
Sources of feed supply

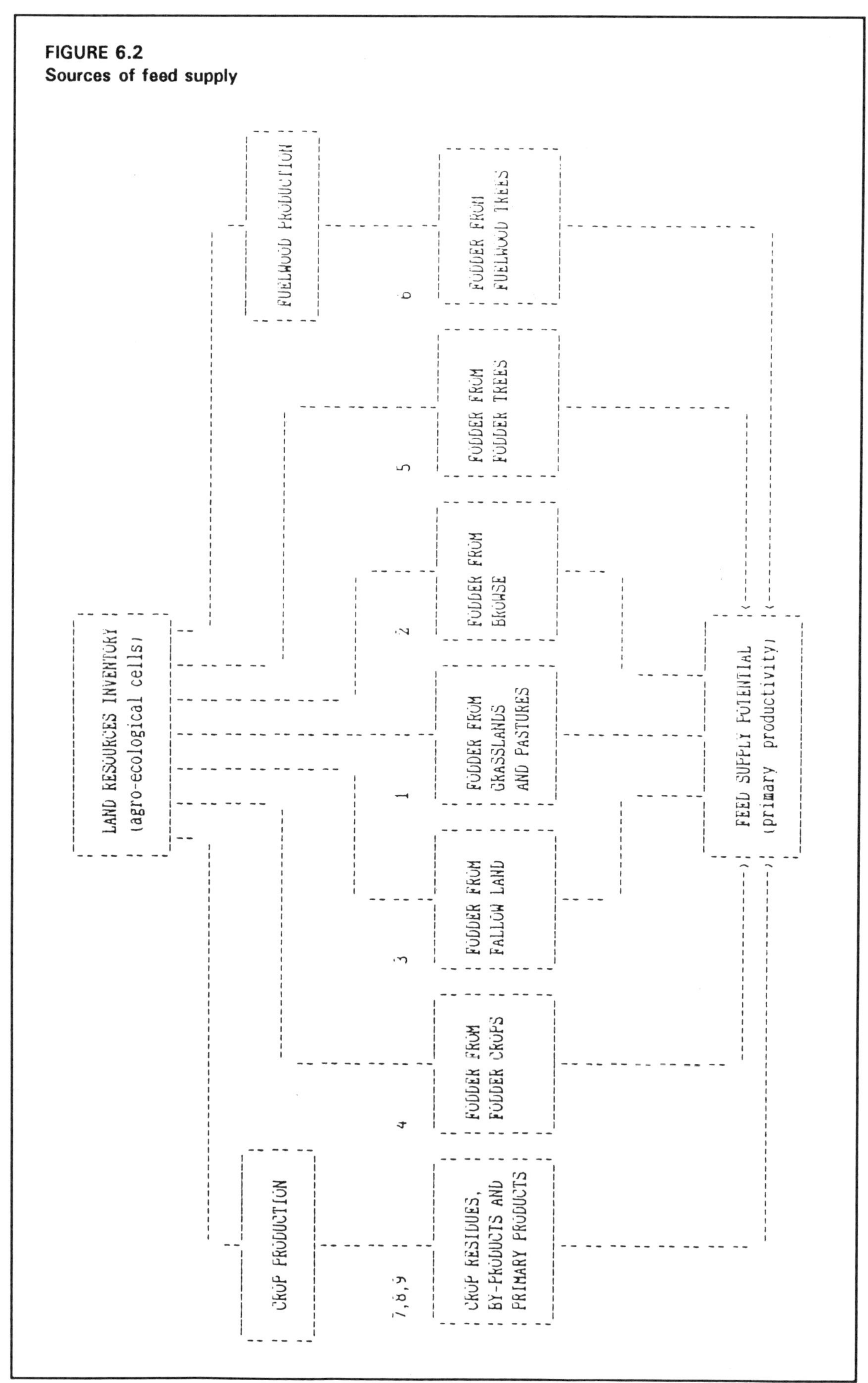

(iii) Fodder from bush and managed fallows within crop rotations (natural or sown grass-legume mixture).

(iv) Fodder crops or fodder grasses, legumes and cereals (sown).

(v) Fodder trees (sown).

(vi) Fodder from fuelwood trees (sown).

(vii) Crop residues.

(viii) Crop by-products.

(ix) Crop primary products.

Feed supplies from each of these sources are described hereunder.

6.1.2 Grasslands and Pastures

Biomass production potential from grasslands or pastures in the model is estimated using the FAO-AEZ method (FAO 1978-81), and involves the following activities (Figure 6.3):

(i) Selection of the species and definition of land utilization types (LUTs) (e.g. species; produce; technology and input level; labour; capital; markets).

(ii) Determination of climatic requirements of species and LUTs and matching climatic requirements with the characteristics of the inventoried climatic zones (thermal zones and growing period zones), and quantifying the climatically attainable yield potentials.

(iii) Determination of edaphic (soil) requirements of species and LUTs, and matching edaphic requirements with the characteristics of the inventoried soil units, textures, phases and stoniness to rate edaphic limitations.

(iv) Quantifying soil erosion hazards (soil loss) in each climate-soil unit (agro-ecological cell) of the land resources inventory by LUT and the associated productivity losses.

(v) Modifying the climatic yield potentials (in ii) according to the soil limitations (in iii) and erosion hazards (in iv) to quantify yield potentials with constraints and ecological land suitabilities of each inventoried climate-soil land unit for each LUT.

Each of these activities is described in the following sections.

6.1.2.1 Land Utilization Types

Considerable work has been done on screening pasture (and fodder grass and legume species) to determine those best suited to particular environments in Kenya (Edwards and Bogdan 1951; Rattray 1960; Pratt and Gwynne 1977; Boonman 1979; Jaetzhold and Schmidt 1982). Table 6.1 sets out a list of some of these pasture (and fodder species) that are considered suitable in Kenya.

FIGURE 6.3
Schematic presentation of suitability assessment for grassland/pasture

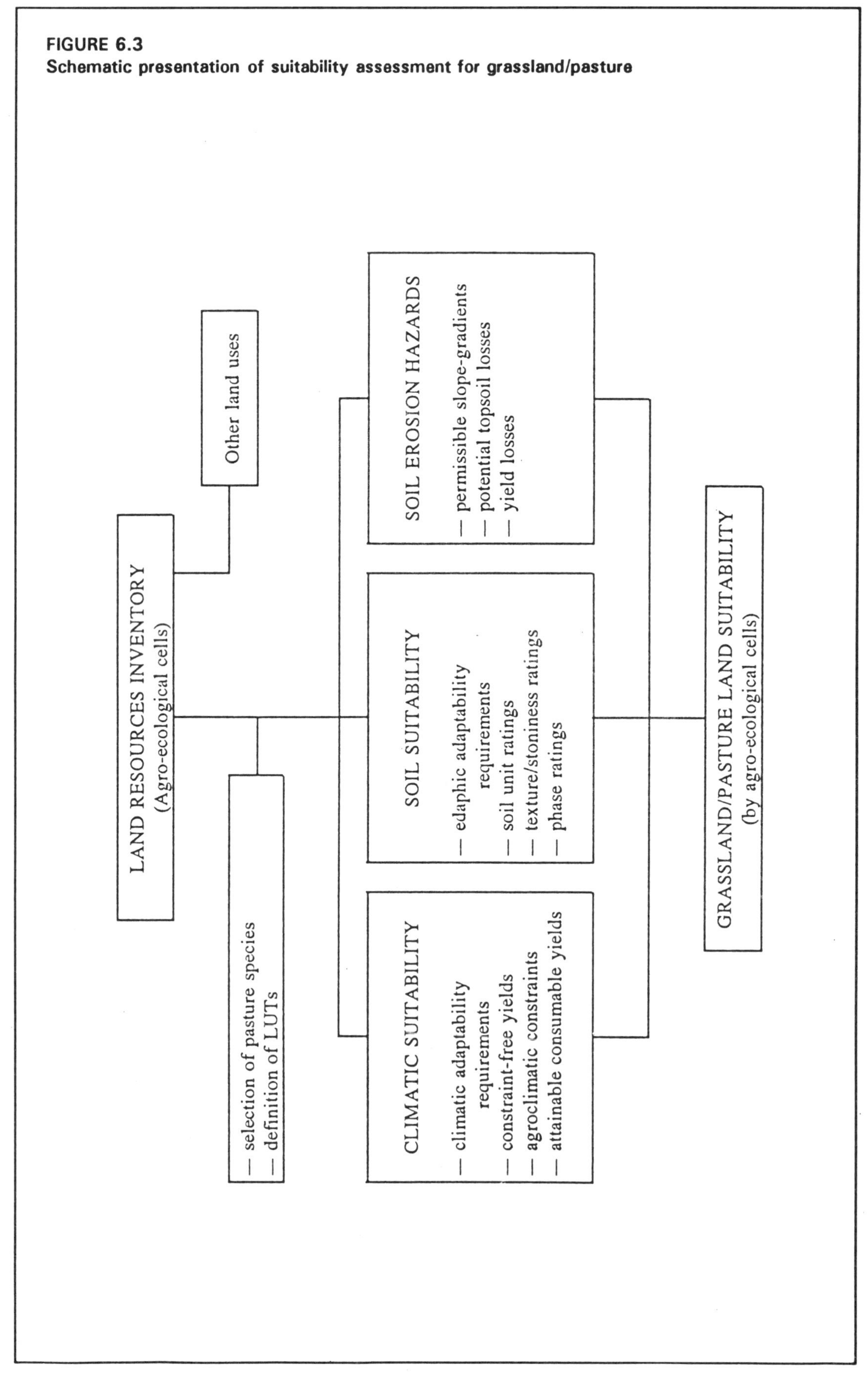

TABLE 6.1
Pasture and fodder species of grasses and legumes

Species name		Common name
Grasses:		
1	*Aristida* spp.	
2	*Cenchrus ciliaris*	Buffel grass
3	*Chloris gayana*	Rhodes grass
4	*Cynodon dactylon*	Star grass
5	*Dactylus glomerata*	Cocksfoot
6	*Digitaria* spp.	
7	*Eragrostis superba*	
8	*Exotheca abyssinica*[1]	
9	*Fescue* spp.	Fescue grass
10	*Hyperrhenia* spp.[1]	Zebra grass
11	*Lolium* spp.	Rye grass
12	*Melinis minutiflora*	
13	*Panicum coloratum*	Coloured Guinea grass
14	*Panicum maximum*	Guinea grass
15	*Pennisetum catabasis*[1]	
16	*Pennisetum clandestinum*	Kikuyu grass
17	*Pennisetum purpureum*[2]	Napier grass
18	*Pennisetum schimperi*	
19	*Setaria spacelata*	
20	*Setaria splendida*[2]	Giant setaria
21	*Sorghum sudanense*[2]	Sudan grass
22	*Sporobolus helvolus*[1]	
23	*Themeda triandra*	Red oat grass
24	*Tripsacum laxacum*[2]	Guatemala grass
Legumes:		
25	*Centrosema pubescens*[2]	
26	*Desmodium* spp.[3]	Tick clover
27	*Lablab purpureus*[2]	Hyacinth bean
28	*Medicago sativa*[2]	Lucerne or alfalfa
29	*Macroptilium atropurpureum*[3]	Siratro
30	*Stylosanthes* spp.	Stylo
31	*Trifolium* spp.	Clover
32	*Vigna* spp.[2]	

[1] Common in areas of impede drainage or in seasonally waterlogged areas.
[2] Fodder species.
[3] Includes pasture and fodder types.

Pasture production is considered at three levels of inputs. The attributes of the three levels of inputs circumstances are listed in Table 6.2 and they form the basis of the definition of land utilization types considered in the model.

6.1.2.2 Climatic Adaptability and Suitability

Climatic suitability assessment of grass and legume pasture species is based on the climatic adaptability principles described in FAO (1978-81), and includes:

(a) an understanding of the climatic adaptability of pasture species in terms of their ecophysiological characteristics;

TABLE 6.2
Attributes of land utilization types considered for pasture and fodder[1] production

Attributes	Low inputs	Intermediate inputs	High inputs
Primary resource	Natural vegetation	Eradication of unpalatable species. Legumes introduced into grassland in zones with >30 days growing period together with selective clearing. Fodder crops and legume sod seeding introduced in zones with >90 days growing period.	Eradication of unpalatable species. Legumes introduced into grassland in zones with >30 days growing period together with selective clearing. Fodder cultivation and legume sod seeding in zones with >90 days growing period.
Water	Surface water, shallow well (lifted by hand)	Boreholes and deep lined wells (windmill/engine driven pumps).	Adequate water reticulation (windmill/engine: gravity distribution).
Land use and feeding system	Traditional (extensive): permanent grazing; no fire control	Controlled (semi-intensive): group ranching; rotational grazing + stall feeding; fire partially controlled; fodder conservation for dry season.	Controlled (intensive): control of numbers and movement; rotational semi-zero and zero grazing; fire controlled; fodder conservation for dry season.
Fertilization	None	In conjunction with legume introduction in region with >90 days growing period application of ground rock phosphate and thiobacillus or superphosphate. With fodder crops, intermediate levels of plant nutrients from fertilizer and manure.	Optimum amount plant nutrients.
Herding	Traditional	Organized	Use of hedges and fences.

[1] Fodder grasses and legumes.

(b) matching the climatic requirements to thermal and moisture regimes, including the estimation of constraint-free biomass potentials;

(c) rating of agro-climatic constraints of water stress/excess; pests, diseases and weeds; and workability; and

(d) estimating attainable biomass production potentials with constraints, and the consumable biomass fractions.

The list of grass and legume species given in Table 6.1 includes both C_4 species (grasses) and C_3 species (legumes). Both groups of species include ecotypes that are adapted to operate under warmer (mean daily temperature > 20°C) as well as cooler (mean daily temperature <20°C) conditions. Table 6.3 presents the thermal zone screen for pasture and fodder species showing which species can be considered in which thermal zones.

TABLE 6.3
Thermal zone screen for pasture and fodder species

	Species name	Adaptability group	Thermal zone							
			T1	T2	T3	T4	T5	T6	T7	T8
Grasses:										
1	*Aristida* spp.	III	●	●	●					
2	*Cenchrus ciliaris*	III	●	●	●	●				
3	*Chloris gayana*	III, IV	●	●	●	●	●			
4	*Cynodon dactylon*	III, IV	●	●	●	●	●	●		
5	*Dactylis glomerata*	IV					●	●	●	●
6	*Digitaria* spp.	III	●	●	●	●				
7	*Eragrostis superba*	III	●	●	●	●				
8	*Exotheca abyssinica*	IV			●	●	●	●		
9	*Festuca* spp.	IV					●	●	●	●
10	*Hyperrhenia* spp.	III, IV	●	●	●	●	●	●		
11	*Lolium* spp.	IV					●	●	●	●
12	*Melinis minutiflora*	IV	●		●	●	●	●		
13	*Panicum coloratum*	III, IV	●	●	●	●	●			
14	*Panicum maximum*	III, IV		●	●	●	●			
15	*Pennisetum catabasis*	IV			●	●	●	●		
16	*Pennisetum clandestinum*	IV			●	●	●	●		
17	*Pennisetum purpureum*	III, IV		●	●	●	●			
18	*Pennisetum schimperi*	IV			●	●	●	●	●	●
19	*Setaria sphacelata*	IV			●	●	●	●		
20	*Setaria splendida*	III, IV	●	●	●	●	●			
21	*Sorghum sudanense*	III, IV	●	●	●	●	●			
22	*Sporobolus helvolus*	III	●	●	●					
23	*Themeda triandra*	IV			●	●	●	●	●	●
24	*Tripsacum laxacum*	III, IV	●	●	●	●	●			
Legumes:										
25	*Centrosema pubescens*	II	●	●	●					
26	*Desmodium* spp.	II	●	●	●					
27	*Lablab purpureus*	II	●	●	●					
28	*Medicago sativa*	I				●	●	●	●	
29	*Macroptilium atropurpureum*	II	●	●	●					
30	*Stylosanthes* spp.	II	●	●	●	●				
31	*Trifolium* spp.	I				●	●	●	●	
32	*Vigna* spp.	II	●	●	●					

Legume species adapted to operate under cool temperatures (<20°C mean daily temperature) belong to adaptability group I, and those adapted to operate under warm temperatures (>20°C) belong to adaptability group II. Grass species adapted to operate under warm temperatures (>20°C) belong to adaptability group III, and those adapted to operate under cool temperatures (<20°C) belong to adaptability group IV. The relationships between photosynthesis and temperature for the four adaptability groups are presented in Table 6.4.

Consequently, the suggested thermal zone combination ratings for the grasses and legumes at this stage of the model development are: S1 (very suitable) for the thermal zones T1, T2, T3, T4 and T5, S2 (suitable) for T6, S3 (moderately suitable) for T7, S4 (marginally suitable) for T8 and N (not suitable) for T9. A rating of S1 indicates that there are no thermal constraints during the growing period and the requirements are fully met. A rating

TABLE 6.4
Relationships between temperature and rate of leaf photosynthesis (kg CH_2O/ha/hr) for legume species in adaptability groups I & II, and grass species in adaptability groups III & IV

Adaptability group	Temperature (°C)							
	5	10	15	20	25	30	35	40
Legume I	2.5	10.0	20.0	25.0	25.0	20.0	10.0	5.0
Legume II	-	2.5	15.0	35.0	37.5	37.5	30.0	20.0
Grass III	-	2.5	30.0	40.0	50.0	50.0	47.5	40.0
Grass IV	2.5	15.0	37.5	50.0	50.0	37.5	25.0	10.0

of S2 indicates slight to moderate thermal constraints leading to yield suppressions of some 25 %. A rating of S3 indicates that there are moderate to severe thermal constraints leading to yield suppressions of some 50%. A rating of S4 indicates severe thermal constraints leading to yield suppressions of some 75%. A rating of N indicates that the thermal requirements are not met and the zone is not suitable for further consideration. These ratings correspond closely with the correlation between pasture dry matter production and temperature in Kenya (Booneman 1979).

Potential biomass estimates were derived according to the method developed by the FAO-AEZ project (Kassam 1977; FAO 1978-81). For the purpose of computing 'constraint-free' biomass potential, it is assumed that both C_3 and C_4 pasture and fodder species of grasses and legumes will be represented, so that maximum photosynthesis rate (Pm) of 37.5 $CH_2O\ ha^{-1}\ hr^{-1}$ has been applied.

Estimates of constraint-free total biomass (Bn) at high level of inputs are presented in Table 6.5 together with maximum leaf area index (LAI) values used, the agro-climatic constraint ratings, total biomass with constraints (Bnc), consumable coefficients (Cc) and consumable biomass with constraints (Bcc).

For a variety of reasons only a portion of the plant biomass is eaten by animals. About 20% of the total net biomass (Bn) is in roots, a portion of the biomass is not eaten (particularly under low inputs) due to low palatability; some biomass is lost due to trampling, fire and wind, and part is consumed by invertebrate animals. It is generally assumed that between a third and two-thirds of the total biomass yield of an area will be utilized or consumed by stock, depending on the environment.

Agro-climatic constraints applied to the constraint-free yield relate to water stress in LGP < 210 days and workability in LGP zone 365^+ days. Total biomass yield with constraint at high inputs range from 0.5 t/ha in LGP zone 1-29 days to 30.6 t/ha in LGP zone 330-364 days.

Consumable coefficients (Cc) range from 0.35 at low inputs in LGP zones with < 120 days to 0.6 at high inputs in LGP zones with > 180 days. Consequently, consumable biomass with constraints (Bcc) range from 0.23 t/ha to 18.9 t/ha for high inputs, from 0.16 to 10.83 t/ha for intermediate inputs, and from 0.09 t/ha to 3.94 t/ha for low inputs.

Constraint-free net biomass (Bn) at low inputs is assumed to be 25% and 50% of those at high input level in the LGP zones > 90 days and < 90 days respectively. Constraint-free total biomass at intermediate level is assumed to be between the high and the low input

TABLE 6.5
Potential biomass from pasture and fodder grasses and legumes (t/ha dry weight) at three input levels[1]

LGP (days)	1-29	30-59	60-89	90-119	120-149	150-179	180-209	210-239	240-269	270-299	300-329	330-364	365^-	365^+
LAI	1-2	2-3	3-4	4	4	4	4	4	4	4	4	4	4	4
Bn	0.0-1.8	1.8-5.3	5.3-10.0	10.0-14.3	14.3-17.2	17.2-19.7	19.7-22.1	22.1-24.3	24.3-26.3	26.3-28.2	28.2-29.7	29.7-31.5	31.5	31.5
'a'[1]	2	2	2	1	1	1	1	0	0	0	0	0	0	0
'b'	0	0	0	0	0	0	0	0	0	0	0	0	0	0
'c'	0	0	0	0	0	0	0	0	0	0	0	0	0	0
'd'	0	0	0	0	0	0	0	0	0	0	0	0	0	1
Bnc	0.5	1.8	3.8	9.1	11.8	13.8	15.7	23.2	25.3	27.3	29.0	30.6	31.5	23.6
Cc – H[2]	0.45	0.45	0.45	0.45	0.50	0.50	0.60	0.60	0.60	0.60	0.60	0.60	0.60	0.60
– I	0.40	0.40	0.40	0.40	0.45	0.45	0.55	0.55	0.55	0.55	0.55	0.55	0.55	0.55
– L	0.35	0.35	0.35	0.35	0.40	0.40	0.50	0.50	0.50	0.50	0.50	0.50	0.50	0.50
Bcc – H	0.23	0.81	1.71	4.10	5.90	6.80	9.42	13.92	15.18	16.38	17.40	18.36	18.90	14.16
– I	0.16	0.54	1.14	2.28	3.32	3.88	5.40	7.98	8.70	9.38	9.97	10.52	10.83	8.07
– L	0.09	0.32	0.67	0.80	1.18	1.38	1.96	2.90	3.16	3.41	3.63	3.82	3.94	2.96

1 Agroclimatic constraints: a = water stress or excess; b = pests, diseases or weeds affecting vegetative growth; c = pests, diseases or weeds affecting reproductive growth; d = workability limitations.

2 H = high inputs; I = intermediate inputs; L = low inputs.

levels. For areas with no growing period Bc is estimated at 70 kg/ha at the high inputs level and 35 kg/ha at the low and 52.5 kg/ha at the intermediate inputs level.

Pasture species and biomass potentials are matched to individual component length of growing periods, i.e. L1, $L2_1$, $L2_2$, $L3_1$, $L3_2$, $L3_3$, $L4_1$, $L4_2$, $L4_3$, $L4_4$. The LGP-Pattern evaluation for pasture is achieved by taking into account all the constituent component lengths in each LGP-Pattern, thus taking into account the year-to-year variability in the number of LGPs per year.

Yields in Table 6.5 apply to normal lengths of growing periods. For intermediate growing periods, yield reductions are of the order of 50% on all soils except Fluvisols and Gleysols. The percentage of occurrence of intermediate lengths of growing periods in all LGP-Pattern zones combined is 100% in LGP zone 1-29 days; 65% in zone 30-59 days; 25% in zone 60-89 days; 10% in zone 90-119 days and 5% in zone 120-149 days.

An exception to the general methodology for climatic suitability assessment applies to areas occupied by Fluvisols because the inventoried length of growing period does not fully reflect their particular circumstances with regard to moisture regime (Section 5.1.2.3). Fluvisols ratings are presented in Technical Annex 5 for the three levels of inputs circumstances.

6.1.2.3 Edaphic Suitability

In order to assess soil suitability for pasture production, the soil requirements of pasture species must be determined. Further, these requirements must be understood within the context of limitations imposed by landform and other features (e.g. soil phases, stoniness) which do not form part of soil composition but have a significant influence on the use that can be made of the soil.

Basic soil requirements for pasture species relate to the following internal soil properties described in Section 5.1.3.2. From the basic soil requirements for pasture species, a number of responses related soil characteristics have been derived. The correlation between the basic soil requirements and soil characteristics given in Table 5.9 has been used to rate pasture and fodder crop performance.

Also as explained earlier (Section 3.2.5), the soil units (Table 3.16) have been defined in terms of measurable and observable properties of the soil itself, and specific clusters of such properties are combined into 'diagnostic horizons ' and 'diagnostic properties '. They are also used to rate soil suitability.

The edaphic suitability classification is input-specific and based on:

(i) matching the soil requirements of pasture and fodder species with the soil conditions of the soil units described in the soil inventory (soil unit evaluation); and

(ii) modification of the soil unit evaluation by limitation, imposed by texture, phase and slope conditions.

The soil unit evaluation for pasture and fodder production is expressed in terms of ratings based on how far the soil conditions of a soil unit meet the growth and production requirements under a specified level of inputs. The appraisal is effected in five basic classes for pasture and fodder grasses and legumes as a group, i.e. very suitable (Sl), suitable (S2), moderately suitable (S3), marginally suitable (S4), and not suitable (N).

A rating of S1 indicates that the soil conditions are optimal, and that suppression of potential yields (if any) is assumed to be nil or slight. A rating of S2 indicates that there are slight to moderate soil constraints and there would be a suppression of potential yields of the order of 25%. A rating of S3 indicates that there are moderate to severe soil constraints and there would a suppression of potential yields of the order of 50%. A rating of S4 indicates that there are severe soil constraints and there would be a suppression of potential yields of the order of 75%. A rating of N indicates that soil conditions are not suitable for production.

The soil unit ratings are presented in Technical Annex 5 and apply as indicated provided there are no additional limitations imposed by soil texture, phase and stoniness. Modifications are required where such limitations are present.

In the case of soil texture, soil unit ratings remain unchanged if the soil is an albic, cambic, ferralic, calcaro-cambic or luvic Arenosol (Q, Qa, Qc, Qf, Qkc, Ql) or a vitric Andosol (Tv), or where textures are medium (fine sandy loam, sandy loam, loam, sandy clay loam, clay loam, silty clay loam, silt), or fine (sandy clay, silty clay, peaty clay, clay). In all other cases (i.e. with coarse textures: sand, loamy coarse sand, fine sand, loamy fine sand, loamy sand) the soil unit rating is one class (25%) lower.

Limitations imposed by phase and stoniness are rated using the five basic classes already described. The stoniness and phase ratings are presented in Technical Annex 5.

6.1.2.4 Slope Limitations and Soil Erosion

Limitations imposed by slope are taken into account in three steps (Chapter 4). Step one defines the slopes which are permissible for pasture production, and as a model variable this is defined as slopes less than 45% (Table 4.1).

Step two involves the computation of potential topsoil loss which is estimated, by input levels, through a modified Universal Soil Loss Equation (Wischmeier and Smith 1978).

Step three relates the estimated topsoil losses to yield losses through a set of equations in Table 4.4, taking into account soil susceptibility (Table 4.3), level of inputs and regeneration capacity of topsoil (Table 4.2).

6.1.2.5 Land Suitability Assessment

All three assessments: the climatic suitability, the edaphic suitability and the soil erosion hazard, are required to determine the ecological land suitability for grassland/pasture production of each climate-soil unit of the land resources inventory. In essence the land suitability assessment takes account of all the inventoried attributes of land and compares them with the requirements of pasture species, to give an easy to understand picture of the suitability of land for grassland/ pasture production.

The results of the land suitability assessment are presented in five basic suitability classes, each linked to attainable yields for the three levels of inputs considered. For each level of inputs, the land suitability classes are: very suitable (VS) - 80% or more of the maximum attainable yield; suitable (S) - 60% to less than 80% of the maximum attainable

FIGURE 6.4
Schematic presentation of the land suitability assessment programme for pasture production

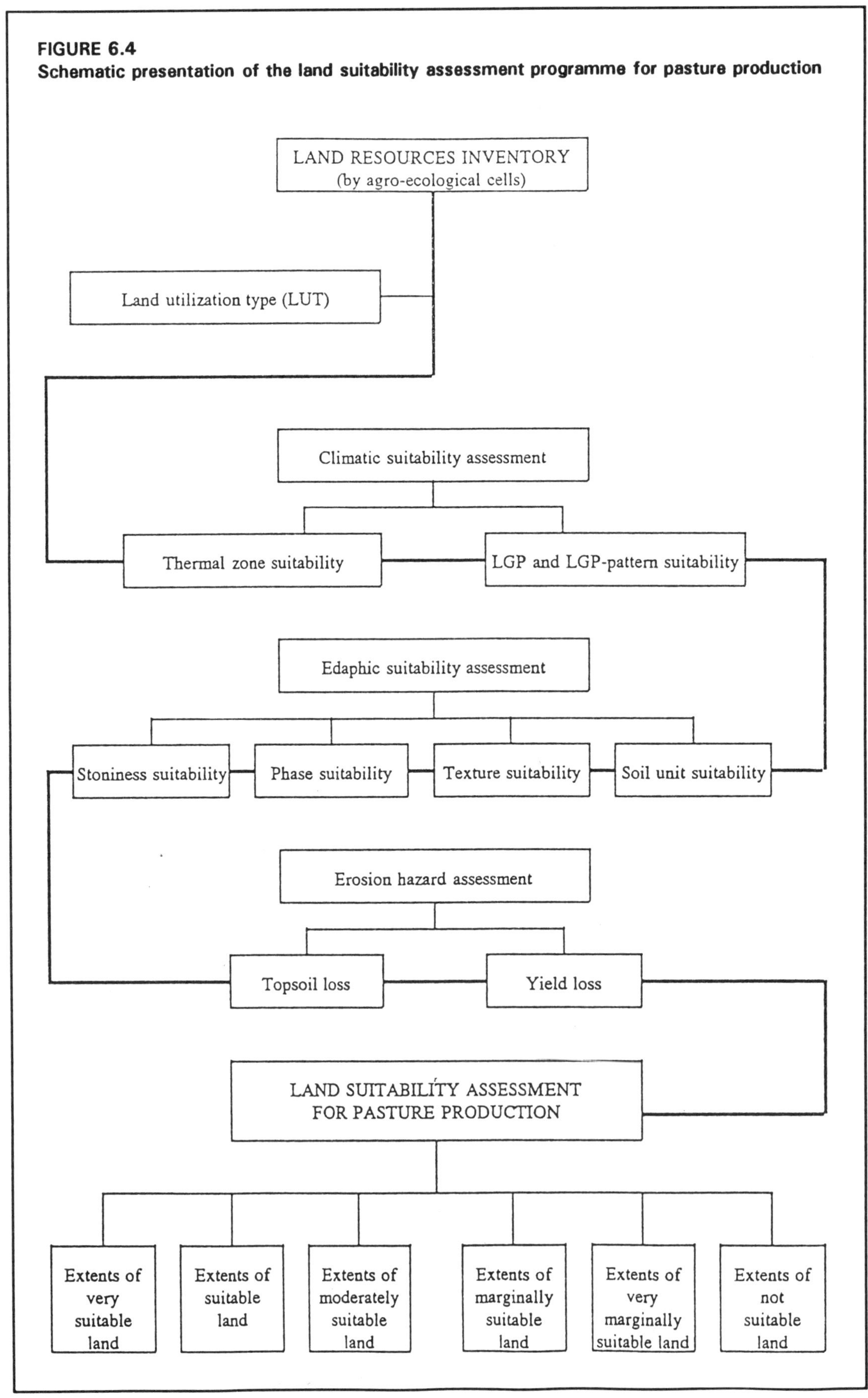

yield; moderately suitable (MS) - 40% to less than 60% of the maximum attainable yields; marginally suitable (mS) - 20% to less than 40%; and not suitable (NS) - less than 20%.

Land suitability assessment is achieved by applying the programme illustrated in Figure 6.4, as explained in Section 5.1.5.

The five classes of land suitabilities are related to attainable yield as a percentage of the maximum attainable under the optimum climatic, edaphic and landform conditions.

Consequently the results provide an assessment of pasture production potentials of each land unit which in turn can be aggregated for any given area in Kenya.

Generalized results of land suitability assessment for pasture production at intermediate level of inputs are presented in Figure 6.5. and in Technical Annex 8. It should be noted that the generalized results presented, include a subdivision of the not suitable class (zero to less than 20% of maximum attainable yield) into two classes (1) very marginally suitable (more than zero to less than 20% of maximum attainable yield) and (2) not suitable (zero yield).

6.1.3 Fodder from Browse, Fodder Trees and Fuelwood Trees

In the low rainfall areas (LGP <120 days), natural woody vegetation including leguminous shrubs and trees, can be important in the nutrition of domestic stock. However, relatively little is known about the digestibility of biomass materials from browse. By comparison with the large amount of herbage from grasslands or natural pastures, the quantity of fodder biomass from natural woody vegetation is limited. Contribution of browse biomass is assumed to be included in the estimates of biomass from grasslands and pastures given in Table 6.5, and no separate account is taken at this stage of the model development and application.

Trees are planted for fodder in Kenya, and the main species are *Acacia, Calliandra, Gliricidia, Grevillea, Leucaena* and *Sesbania*. Again the potential contribution from sown fodder trees is assumed to be included in the estimates of biomass from pastures given in Table 6.5, and no separate account is taken at this stage of model development and application. However, the land suitability procedure for separately quantifying fodder biomass from fodder trees is identical to the procedure of quantifying wood biomass from fuelwood trees in Chapter 7. Consequently, it is now possible, if required, to provide for a seperate assessment of fodder from fodder trees.

Where trees are considered for fuelwood production and carry palatable foliage, it is assumed that about 10% (i.e. 3.3% of mean annual wood biomass increments given in Chapter 7) of the foliage may be utilized by stock without affecting fuelwood yields. Fuelwood species that can contribute fodder are: *Acacia gerrardia, A. nilotica, A. senegal, Calliandra calothyrus, Casuarina equisetifolia, Conocarpus lancifolius, Eucalyptus camaldulensis, E. citridoria, E. tereticornis, Parkinsonia aculeata, Sesbania sesban.*

6.1.4 Fodder from Fallow Land

In the crop productivity model (Chapter 5), fallow requirements for crop rotation options are formulated. At low level of inputs, fallow land is assumed to carry natural bush vegetation; at intermediate and high levels of inputs, fallow land is assumed to carry sown grass-legume pasture.

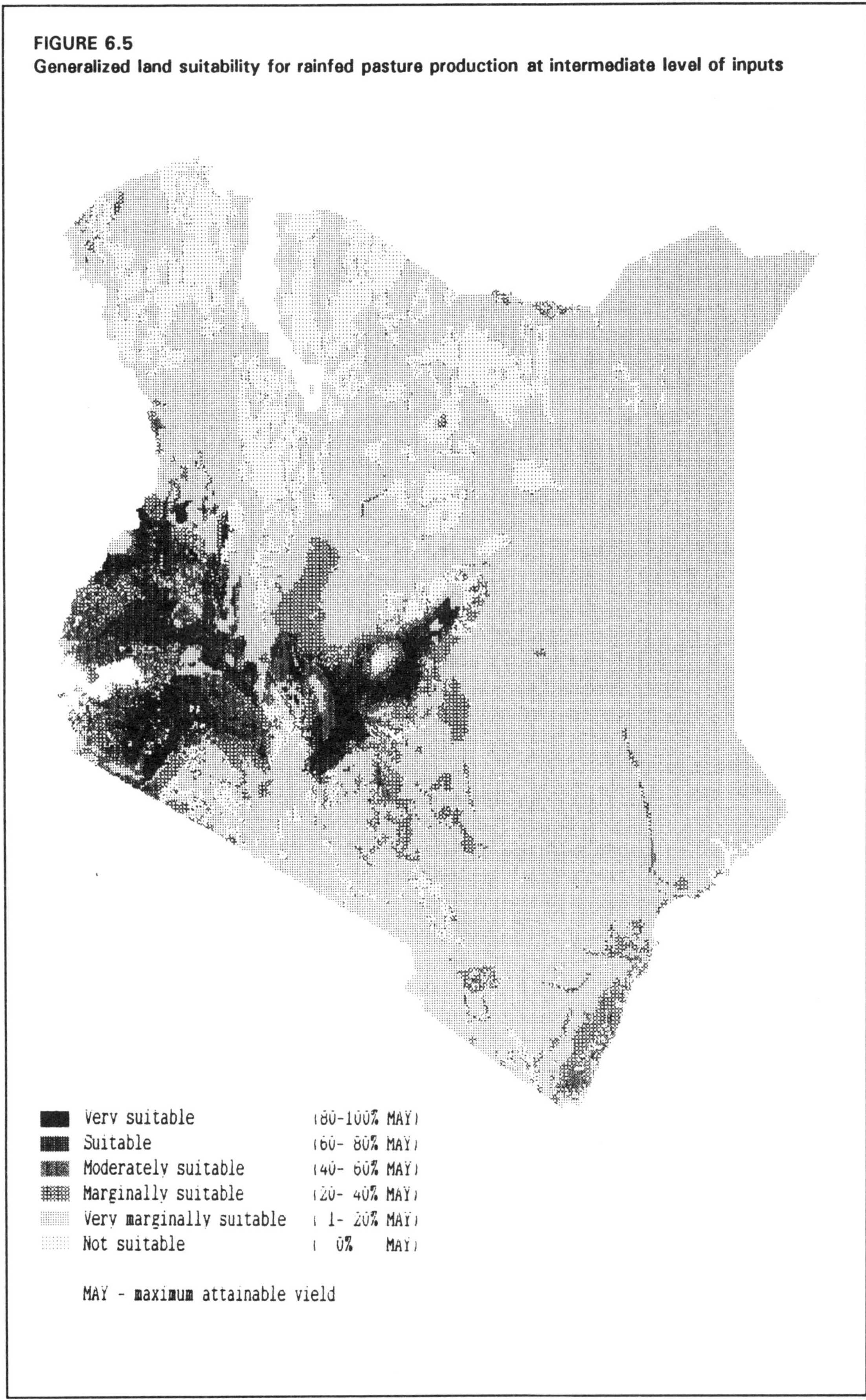

FIGURE 6.5
Generalized land suitability for rainfed pasture production at intermediate level of inputs

Biomass production from natural fallow under low inputs, and from sown pasture under intermediate and high inputs is taken as one-third of that from normal sown or permanent pastures given in Table 6.5. It is further assumed that only 50% of the biomass may be utilized by stock.

6.1.5 Fodder from Fodder Crops

Fodder grasses, legumes and cereals are grown for fodder production in Kenya. Main fodder grasses are *Pennisetum purpureum* (Napier or Bana grass), *Setaria splendida* (Giant setaria), *Sorghum sudanense* (Sudan grass) and *Tripsacum laxacum* (Guatemala grass). Main fodder legume species are *Centrosema pubescense, Dolichos Lablab purpureus* or *Lablab niger* (Hyacinth bean), *Macroptilium atropurpureum* (Siratro), *Vigna* spp. and *Stylosanthes* spp. Main fodder cereals are maize, oat, pearl millet and sorghum.

A separate assessment of biomass potential from fodder grasses, legumes and cereals is possible according to the land suitability methodology described in Chapter 5. However, at this stage of model development and application, the range of biomass potentials from pastures given in Table 6.5, are found to adequately cover the biological potentials of fodder crops.

6.1.6 Crop Residues, By-products and Primary Products

In areas with more than 120 days growing period, crop residues are an important source of fodder particularly for the low and intermediate technology livestock systems. Important residues are the haulms of groundnut, cowpea and other grain legumes, and the staves (stalks) of sorghum, maize and millet, and straw from rice, wheat, barley and oat. Quantities of residues that may be available have been estimated by applying the residue factor and the corresponding utilization coefficients and to crop yields (Technical Annex 5).

By-products, defined as edible materials remaining after a crop has been processed, are bran, pollard and germ meal from cereal milling; molasses and bagasse from sugar milling; and cakes (cotton, soybean, groundnut) from oilseeds. Quantities of crop by-products that may be available have been be estimated by applying the by-product factor and the corresponding utilization coefficients to crop yields (Technical Annex 5).

The term primary product applies to grain used for the purpose of feeding to animals either directly in an unprocessed form or in a processed form. Main cereals used in Kenya are maize, sorghum wheat and barley. Direct grain feeding is used mainly at the high level of technology in the dairy and meat production systems with cattle and goat. The intensive livestock industries of poultry and pig production tend to rely on processed feeds.

6.1.7 Feed Supply Potential (Primary Productivity)

When Part I of the livestock productivity model (Figure 6.1) is applied to the land resources inventory, feed supply potential of each agro-ecological cell are quantified by feed source (Figure 6.2), as described earlier.

Once feed supply potential or primary productivity has been quantified, it is possible to quantify livestock productivity potential of livestock systems at specified performance levels and feed requirements. These aspects are taken into account in Parts II, III, IV and V of the model.

TABLE 6.6
Attributes of the non-pastoral land utilization types considered for livestock production

Attributes	Low inputs	Intermediate inputs	High inputs
Nutrition	Traditional	Mineral supplements, improved calf care, better use of residues and products	As intermediate, plus feeding for optimum economic and biological production; use of primary products
Disease control	None	Current veterinary prophylaxis, plus control or eradication of of diseases or their vectors, e.g., dipping against ticks	As intermediate, plus control of diseases of high performance, e.g., dipping and drenching for mastitis, foot-and-mouth, etc.
Breeding	Selection of unrelated bulls of good conformation, minimum size of heifer at mating	Introduction of adapted exotic breeds and crossbreds	Introduction of graded and exotic breeds of high genetic potential for growth and milk production
Marketing	Low off-take, poor transport facilities, poor processing, including hides and skins	Better off-take, transport and processing, organized markets	Stratified livestock industries, producers receive fair price, organized dairying

6.2 Characterization of Livestock Systems

Part II of the livestock productivity model (Figure 6.1) characterizes the livestock systems that are to be considered in assessing secondary productivity potentials. It defines, for three levels of input situations (or technology levels), the livestock types, production systems and herd structures.

Of the six types of livestock which are considered in the model, four are considered under pastoral as well as non-pastoral systems. They are: cattle, goat, sheep, camel. The remaining two, poultry and pig, are considered under intensive systems only and without explicitly defining the production systems at this stage in the model development.

Cattle, goat, sheep and camel systems are considered at three inputs levels. The attributes of the three input level production circumstances for non-pastoral systems are presented in Table 6.6 and form the basis of the definition of the non-pastoral utilization types considered in the model.

For the pastoral systems, three types of cattle herds have been considered. These are: nomadic distant, nomadic with market accesss and semi-nomadic, representing respectively the low, intermediate and high level of inputs circumstances. For sheep and goat, two types of herds have been considered. These are: nomadic distant and semi-nomadic, representing respectively the low and high level of inputs circumstances. For camel, one herd type has been considered, representing the normal circumstances of production at a low level of inputs.

Herd structures have been defined in terms of number of heads of animal as well as in terms of reference Tropical Livestock Unit (TLU) defined as a mature animal weighing 250 kg (Houerou and Hoste 1977; Stotz 1983).

Livestock conversion factors for non-pastoral systems in areas with more than 120 days growing period are taken from Stotz (1983). Livestock conversion factors for pastoral systems in areas with less than 120 days growing period are taken from Houerou and Hoste (1977), and are:

Cattle in Herd	= 0.70 TLU	Goat	= 0.08 TLU
Cow	= 1.00 TLU	Donkey	= 0.50 TLU
Sheep	= 0.10 TLU	Camel	= 1.25 TLU

6.2.1 Cattle Systems: Dairy and Meat

At the low level of technology, the systems are characterized by pure Zebu cattle (Stotz 1983).

The feed supply is generally native Kikuyu/star grass pastures, and crop residue (maize stover). Cattle are grazed, herded or tethered during the day and kraaled during the night. Cattle are not supplied with concentrates or mineral supplements.

Calves join their dams during milking and for a short while afterwards, during which time they consume the remaining milk in the udder amounting to about 3 to 5 litres per day (400 litres total during the rearing period). Calves are weaned about 5 to 7 months old.

The animals are driven to water at rivers or reservoirs twice a day if nearby, otherwise once a day. Disease control measures are rarely practised, but cattle are compulsorily vaccinated against rinderpest and in some areas against foot and mouth disease.

At the intermediate technology level the cattle would be first generation crossbreds with exotic or high performing grade cattle bred with the help of artificial insemination. Crossbred cattle are generally acquired through upgrading local Zebu cows by Ayrshire, Friesian, Guernsey or Jersey bulls.

Cattle graze natural ley pastures (Kikuyu/star grass) and fields are usually fenced. With regards to feeding, young stock rearing and watering, the same husbandry practice is employed as for Zebu cattle. Crossbred cattle occasionally receive cattle salt, and cattle are regularly dipped or sprayed. Sick animals are treated.

Under the high inputs situations, the systems are based on exotic cattle, Friesian, Ayrshire, Guersey or Jersey, which have been 'graded ' up from the original crosses with indigenous cattle bred back to the exotic type.

Stotz (1983) describes these systems as charaterized by grade cows in a combined grazing/stall feeding system (semi-zero grazing) or complete stall feeding (zero grazing). In the case of the semi zero grazing, cattle usually graze Kikuyu or star grass during the daytime. At night cattle are kraaled or stabled where they are fed with napier and bana grass. Sometimes they are also fed during the day with crop residues and napier grass, particularly during the dry season when pasture productivity is low. Where cattle is permanently housed

TABLE 6.7
Cattle herd structures

Parameter	Input levels					
	Low		Intermediate		High	
Body weight (kg)	Weight	TLU	Weight	TLU	Weight	TLU
Cows	250	1.00	300	1.25	400	1.75
Replacement heifers (ave. of 1 & 2 yr old)	190	0.76	235	0.94	315	1.26
Calf birth weight	22	0.09	25	0.14	35	0.14
Weaning weight (6 months)	80	0.32	100	0.40	140	0.56
Bulls	300	1.20	-	-	-	-
Number in herds	Head	TLU	Head	TLU	Head	TLU
Cows						
- in milk	67	67	72	95	85	149
- dry	33	33	28	30	15	26
- total	100	100	100	100	100	175
Calves						
- heifers	33	-	36	-	42	-
- bulls	34	-	36	-	(43)	-
- total	67	23.5	72	30.4	(85)42	25.8
Replacement						
- heifers	29	22.0	37	32.0	51	63.8
- bulls	2	1.5	-	-	-	-
- total	31[1]	23.5	37	32.0	51	63.8
Bulls	4	4.8	-	-	-	-
Total	202	151.8	209	187.4	194	264.6

1 Contribution to meat output accounted for in FAO/IIASA (1991:Tech. Annex 5, Table A6.2).

TABLE 6.8
Goat herd structures

Numbers in herd	Input levels		
	Low	Intermediate	High
Breeding does	100	100	100
Kids (under 6 months)	127	156	187
Replacement yearlings			
- female	19	25	19
- male	2	2	2
Bucks	4	4	4
Total head	252	287	312
Total adult animals	125	131	125
TLU/adult head	0.10	0.11	0.12
Total TLU	12.5	14.4	15.0

in a shed, the feed is cut and carried to them. Cattle kept in a zero grazing unit are predominantly fed with napier or bana grass which is first chopped. Dairy cows also receive 20-25 kg/cow per year of mineral supplement, and 500 to 1000 kg/cow per year manufactured and compound concentrates when lactating.

Male and female calves are bucket fed and hand reared. Calves usually receive 270 to 400 litres of milk only until they are weaned within 10 to 18 weeks. When the weaning period is shorter in the case of zero grazing system, calves also receive about 165 kg of concentrates during the rearing period. After this time they depend entirely on forage and join the rest of the herd about 6 months old at a weight of about 160 kg, or at two weeks at a weight of about 35 kg in the zero-grazing system.

Animals are watered twice a day and are regularly dipped or sprayed, drenched against internal parasites and receive other health treatment as needed.

Herd structures parameters are presented for the three cattle herd types in Table 6.7. Base herd structures are defined on the basis of a notional herd of 100 cows.

6.2.2 Goat Systems: Dairy and Meat

Under the low input technology, the system is characterized by the local small East African goats which are herded or tethered during the day and kept in store, stable or some kind of shelter at night. Goats feed mainly on natural pasture which supply about 70% of all feed consumed. The remainder is obtained for crop residues and through browsing on farm hedges. There is no definite mating season, hence kids are born the whole year round. Kids suckle the mother for about 5 to 7 months and consume the whole amount of milk produced by the dam.

Under the intermediate technology level, the systems are characterized by the dual purpose goats, usually F1 or F2 cross breeds. Generally, an exotic dairy goat buck such as Toggenburg, Saanen or Anglo-Nubian, is used for upgrading local goats. Animals are kept under semi-zero grazing management system. They are tethered during the day, graze mainly natural pasture, and frequently browse shrubs and farm hedges. Goats are penned during the night when they are fed with crop residues and fodder crops like napier grass and maize. Under these feeding conditions, goats obtain approximately 40% of their dry matter requirements from grazing natural pastures. Another 40% is drawn from fodder crops and 20% supplied through feeding crop residues. Animals are sprayed with acaricides regularly and drenched against internal parasites at regular intervals. Lactating females are partially milked before kids are allowed to suckle with an off-take of 1 to 2 kg of milk daily.

The high level of technology situation is characterized by the intensive goat production system. The main aim of keeping exotic or grade dairy goats like Toggenburg, Saanen or Anglo-Nubian is to produce milk. Other by-products are sales of breeding stock and goat meat. Goats are usually kept in a zero grazing system, where they are fed with napier grass and other fodder crops along with upto 1.5 kg of concentrate per day. Water is provided in containers which are placed inside the stable. Kids are bucket fed with milk, obtain 165 litres over a period of 4 months and are supplemented with 50 kg of concentrates. All animals are sprayed with acaricide regularly.

Herd structure parameters are presented for the three goat herd types in Table 6.8. Base herd structures are defined on the basis of a notional herd of 100 does.

TABLE 6.9
Sheep herd structures

Numbers in herd	Input levels		
	Low	Intermediate	High
Breeding ewes	100	100	100
Lambs (under 6 months)	115	148	150
Replacement yearlings			
- female	19	22	22
- male	2	2	2
Rams	4	3	3
Total head	240	275	277
Total adult animals	125	127	127
TLU/adult head	0.10	0.10	0.10
Total TLU	12.5	12.7	12.7

TABLE 6.10
Herd proportions (%) by districts of nomadic herds in areas with LGPs <120 days, expressed in TLUs

District	Cattle	Camel	Smallstock	Donkey
Mandera	21.0	65.8	13.0	0.2
Wajir	28.0	64.7	7.0	0.3
Turkana	31.0	29.2	37.5	2.3
Marsabit	54.0	27.5	16.5	2.0
Garissa	76.8	15.6	7.5	0.1
Lamu	76.8	15.6	7.5	0.1
Tana River	66.4	21.0	11.6	1.0
Kiliji	66.4	21.0	11.6	1.0
Isiolo	64.3	16.4	17.5	1.8
Baringo	65.4	12.2	21.9	0.5
Samburu	61.9	6.6	29.3	2.2
Taita Taveta	83.2	-	14.2	2.6
Kwale	83.2	-	14.2	2.6
Kajiado	80.0	-	18.6	1.4
Narok	80.0	-	18.6	1.4

6.2.3 Sheep Systems: Meat and Wool

The dominant local breed kept at the low technology level is the Red Maasai or Red Kikuyu. It is a fat tailed hair sheep weighing some 25 to 30 kg. The animals feed mainly on natural pasture both on the farm and adjacent common land, and use crop residues and other consumable dry matter that can be found. Animals are often tethered during the day and kept in some kind of shelter at night. There is no definite mating season or control over breeding and little health care. Ewes on average lamb once a year. No milk is taken and the only product is meat from surplus male lamb and cull ewes.

At the intermediate level of technology, there is controlled breeding and introduction of better class of sire, usually Droper rams, to improve meat production. The preferred crossbred seems to be 3/4 Droper and 1/4 Maasai. In conjunction with this, there is more frequent joining programme, regular dipping and drenching, improvement in fodder provided and mineral supplementation. These inputs are accompanied by fenced paddocks rather than tethering or shepherding.

At the high level of technology, the production system is characterized by the Red Maasai x Droper crosses for meat production. At higher elevations, (thermal zones T5, T6, T7 and T8), another system dominated by dual purpose wool and meat production, using crossbred wool sheep such as Corriedale-Hampshire, is also considered in the model.

Herd structure parameters are presented for the three sheep herd types in Table 6.9. Base herd structures are defined on the basis of a notional herd of 100 ewes.

6.2.4 Pastoral Systems: Meat and Milk

Pastoral systems have evolved as a method of producing human food under climatic conditions where normal rainfed crop production is not possible. It operates in the semi-arid zones where the rainfall is low in total quantity and is erratic both geographically over land in time, that is within seasons and between seasons.

The system comprises various combination of large and small domesticated ruminants with the variations dictated by climate, notably temperature. As a source of food the large ruminants provide milk and some blood and meat while the small ruminants are a source of meat and, in certain locations, of milk. Camels play no part in the market food economy so they are not managed with a view to producing saleable surplus. The role of the camel is mainly to provide milk and be a beast of burden.

All of the pastoral system operates on various combinations of cattle, camels, sheep and goat with some donkeys as pack animals. The combinations are a function of climate, available herbage and water, and local preferences. In the north the herds are principally camels/smallstock with some cattle in certain locations while in the centre and south the herds are almost exclusively cattle/smallstock. The herd proportions for the principal pastoral districts expressed in TLU equivalent (1 TLU = 250 kg animal) are set out in Table 6.10.

The proposed herd structure for cattle, derived from Unesco (1982), Semenye (1982) and Meadows and White (1981), are presented in Table 6.11. Herd structures for sheep and goat, derived from Unesco (1982), de Leeuw and Peacock (1982), Peacock (1983, 1984) and King, Sayers, Peacock and Kontrohr (1982), are presented in Table 6.12. The proposed herd structure for camel is given in Table 6.13.

6.3 Quantification of Herd Performance

Part III of the livestock productivity model (Figure 6.1) quantifies livestock productivity potential of each livestock system by quantifying herd performance in acceptable climatic zones.

The thermal zone suitablity ratings for livestock systems are given in Table 6.14. The moisture zone screen, indicating which livestock systems can be considered in which growing period zones, is presented in Table 6.15.

Livestock products per reference herd TLU for cattle, goat, sheep and camel systems at low, intermediate and high levels of technology are presented in Table 6.16 for zones that are considered as S1 for these livestock systems. Where a thermal zone rating is S2, S3 or S4, reference output must be decreased by 25%. 50% and 75% respectively. Where a thermal zone rating is N, the zone is either deemed not suitable because of temperature

TABLE 6.11
Pastoral cattle herd[1] structures

Numbers in herd	Semi-nomadic	Nomadic with market access	Nomadic distant
Breeding cows			
- in milk	23	23	23
- dry	13	22	22
- total	36	45	45
Replacement heifers (1 to 4 yr. old)	22	22	22
Heifer calves	10	10	10
Sub-total females	68	77	76
Steers 1-2 yrs	7	5	6
2-4 yrs	13	9	4
Bull calves	8	8	5
Bulls	5	5	6
Sub-total males	33	27	21
Herd total	101	104	97
Total TLU	70.7	72.8	67.9

1 Nomadic distant - low inputs; Nomadic with market access - intermediate inputs; Semi-nomadic - high inputs.

TABLE 6.12
Pastoral sheep and goat herd[1] structures

Numbers in herd	Semi-nomadic	Nomadic distant
Sheep : Goat ratio	1 : 1.2	1 : 1
Sheep		
- Ewes	50	50
- Ewes weaners	18	13
- Ewes lambs	20	16
Sub-total females	88	79
- Ram lambs	20	15
- Wethers	20	24
- Rams	4	4
Sub-total males	44	43
Total Sheep	132	122
Total TLU	13.2	12.2
Goat		
- Doe	54	60
- Weaner does	18	13
- Kid does	24	18
Sub-total females	96	91
- Kid billies	23	18
- Wethers	33	23
- Billies	4	4
Sub-total males	60	45
Total Goat	156	136
Total TLU	12.5	10.2

1 Nomadic distant — low inputs; Semi-nomadic — high inputs.

TABLE 6.13
Pastoral camel herd[1] structure

Numbers in herd	Nomadic
Breeding females (6-13 yrs)	
- in milk	21
- dry	21
- total breeders	42
Breeder replacements (2-6 yrs)	22
Female calves	2
Total females	70
Bull calves	4
Bull replacements (2-4 yrs)	13
Bulls (5-12 yrs)	6
Castrates (5-12 yrs)	12
Total males	35
Total Herd	105
Total TLU	131

1 Nomadic - low inputs.

TABLE 6.14
Suitability ratings for livestock systems by thermal zone

Livestock system	Thermal zone								
	1	2	3	4	5	6	7	8	9
Cattle:									
1 Dairy and meat	S1	S1	S1	S1	S1	S1	S1	S3	N
2 Pastoral	S1	S1	S1	S1	S2	S3	S3	N	N
Goat:									
3 Dairy and meat	S1	S1	S1	S1	S1	S1	S1	S3	N
4 Pastoral	S1	S1	S1	S1	S2	S3	S3	N	N
Sheep:									
5 Meat and wool	S1[1]	S1[1]	S1[1]	S1[1]	S1	S1	S1	S2	N
6 Pastoral	S1	S1	S1	S1	S2	S3	S3	N	N
Camel:									
7 Pastoral	S1	S1	N	N	N	N	N	N	N
Others:									
8 Poultry	S	S	S	S	S	S	S	N	N
9 Pig	S	S	S	S	S	S	S	N	N

1 N for wool production in T1,T2,T3,T4 and S2 in T5.

constraints (and therefore not considered further), or it is deemed not applicable for further consideration because the zone has not been selected for assessment within a particular planning scenario. Where the thermal zone rating is S, as in the case of poultry and pig under intensive system, it represents a screening device to indicate that the zone is deemed suitable for further consideration.

The herd performance parameters calculations for cattle, goat, sheep and camel system are given in Technical Annex 5, and show how output performance values set out in Table 6.16 are derived.

TABLE 6.15
Suitability ratings for livestock systems by LGP zone

Livestock system	Length of growing period zone (days)					
	0	1-29	30-59	60-89	90-119	> 120
Cattle:						
1 Dairy and meat	N	N	N	N	N	S
2 Pastoral	S[1]	S[1]	S[1]	S[1,2]	S[1,2]	N
Goat:						
3 Dairy and meat	N	N	N	N	N	S
4 Pastoral	S[1]	S[1]	S[1,2]	S[1,2]	S[1,2]	N
Sheep:						
5 Meat and wool	N	N	N	N	N	S
6 Pastoral	S[1]	S[1]	S[1,2]	S[1,2]	S[1,2]	N
Camel:						
7 Pastoral	S[1]	S[1]	S[1]	S[1]	N	N
Others:						
8 Poultry	N	N	N	N	S	S
9 Pig	N	N	N	N	S	S

1 Nomadic
2 Semi-nomadic

S — Suitable for consideration
N — Not suitable for consideration

TABLE 6.16
Output of livestock products per herd TLU

Livestock system	Input technology			
	Product	Low	Intermediate	High
Cattle:				
1 Dairy and meat	Milk[1]	264.8	567.8	901.5
	Meat[1]	24.6	27.9	19.8
	Draught[2]	0.09	0.1	0.02
2 Pastoral	Milk	59.3	60.0	67.9
	Meat	15.4	18.6	24.6
Goat:				
3 Dairy and meat	Milk	-	263.7	2166.7
	Meat	92.6	114.6	132.7
4 Pastoral	Meat	7.6	13.7	19.8
Sheep:				
5 Meat and wool	Meat	70.5	123.0	132.2[3]
				107.2[4]
	Wool	-	-	25.0[4]
6 Pastoral	Meat	8.9	14.2	19.4
Camel:				
7 Pastoral	Milk	96.2	120.6	144.3
	Meat	1.9	2.4	2.9

1 Milk in litres; Meat in kg dressed weight; Draught animals in TLUs.
2 Reduce meat output by 45%, 49% and 13% in low, intermediate and high input systems respectively when considering draught animal output.
3 Meat output of 132 kg/TLU applies when there is no wool production.
4 Meat output of 107.2 kg/TLU applies when there is wool production of 25.0 kg/TLU in thermal zones T5, T6, T7, and T8 (Table 60).

6.3.1 Cattle Systems: Dairy and Meat

These systems are considered in thermal zones T1, T2, T3, T4, T5, T6, T7 and T8 (Table 6.14) in growing period zones of more than 120 days (Table 6.15).

For the low technology system, output performance per TLU is 264.8 litres milk and 24.6 kg meat. If draught animals were desired then upto 0.09 TLU of draught animals per TLU could be produced but there would be upto 45% proportional reduction in the meat output (Table 6.16).

For the intermediate technology system, output performance per TLU is 768.4 litres milk and 26.0 kg of meat. If draught animals were desired then up to 0.11 TLU of draught animals per TLU could be produced but there would be up to 49% proportional reduction in the meat output (Table 6.16).

For the high technology system, output performance per TLU is 901.5 litres milk and 19.8 kg meat. If draught animals were desired then up to 0.02 TLU of draught animals per TLU could be produced but there would be up to 13% proportional reduction in the meat output (Table 6.16).

6.3.2 Goat Systems: Dairy and Meat

These systems are considered in the thermal zones T1, T2, T3, T4, T5, T6, T7 and T8 (Table 6.14), and in growing period zones of more than 120 days (Table 6.15).

For the low technology system, output performance per TLU is 92.6 kg meat. For the intermediate technology system, output performance per TLU is 263.7 litres milk and 114.6 kg meat. For the high technology system, output performance per TLU is 2166.7 litres of milk and 132.7 kg meat (Table 6.16).

6.3.3 Sheep Systems: Meat and Wool

These systems are considered in thermal zones T1, T2, T3, T4, T5, T6, T7 and T8 (Table 6.14), and in growing period zones with more than 120 days (Table 6.15).

For the low technology system, output performance per TLU is 70.5 kg of meat. For the intermediate technology system, output performance is 123 kg meat. In the high technology system output performance per TLU is 151.8 kg of meat. In thermal zones 6 and 7, output performance per TLU is 126.9 kg of meat and 25 kg wool (Table 6.16) for the meat and wool system.

6.3.4 Pastoral Systems

6.3.4.1 Cattle: Meat and Milk

These systems are considered in thermal zones T1, T2, T3, T4, T5, T6 and T7 (Table 6.14), and in growing period zones less than 119 days (Table 6.15) except for the semi-nomadic herd (high technology) which is considered only in growing period zones 60-89 days and 90-119 days (Table 6.16).

For the nomadic distant herd (low technology), output performance per TLU is 59.3 litres milk and 15.4 kg meat. For the nomadic herd with market access (intermediate technology), output performance per TLU is 60 litres milk and 18.6 kg meat. For the semi-nomadic herd (high technology), output performance per TLU is 67.9 litres milk and 24.6 kg meat (Table 6.16).

6.3.4.2 Goat: Meat

These systems are considered in thermal zones T1, T2, T3, T4, T5, T6 and T7 (Table 6.14). The nomadic distant herd is considered in growing period zones with less than 120 days. The semi-nomadic herd is considered in growing period zones 30-59 days, 60-89 days and 90-119 days (Table 6.15).

The nomadic distant herd is assumed to represent the low technology system, and its output performance per TLU is 7.6 kg meat. The semi-nomadic herd is assumed to represent the high technology system, and its output performance per TLU is 19.8 kg meat (Table 6.16). The output performance per TLU for the intermediate technology system is assumed to be half-way between the low and the high technology performance (i.e. 13.7 kg meat).

6.3.4.3 Sheep: Meat

These systems are considered in thermal zones T1, T2, T3, T4, T5, T6 and T7 (Table 6.14). The nomadic distant herd is considered in growing period zones with less than 120 days. The semi-nomadic herd is considered in growing period zones 30-59 days, 60-89 days and 90-119 days (Table 6.15).

The nomadic distant herd is assumed to represent the low technology system, with an output performance per TLU of 8.9 kg meat. The semi-nomadic herd is assumed to represent the high technology system, with an output performance per TLU of 19.4 kg meat (Table 6.16). The output performance at the intermediate level is assumed to be half-way between the low and the high technology performance (i.e. 14.2 kg meat).

6.3.4.4 Camel: Meat and Milk

The system is considered in thermal zones T1 and T2 (Table 6.14), and in growing period zones less than 90 days (Table 6.15).

The output performance per TLU for the nomadic herd is 96.2 litre milk and 1.9 kg meat (Table 6.16) This output performance per TLU is assumed to apply at the low inputs level. The output performance per TLU at the high inputs level is assumed to be 50% greater (i.e. 144.3 litre milk and 2.9 kg meat), and the intermediate level performance is assumed to be half-way between the low and the high level performance (i.e. 120.6 litre milk and 2.4 kg meat).

6.3.5 Poultry and Pig: Meat and Egg

Poultry and pig production has been considered to apply only under the intensive system. The feed conversion ratios for poultry meat and eggs and pig meat are given in Section 6.4.5. Performance parameters have not been explicitly formulated for poultry and pig system at this stage of the model development but it is envisaged that these would be incorporated at a later stage.

6.3.6 Pests and Diseases

Major diseases of cattle include rinderpest, trypanosomiasis, contagious bovine pleuropneumonia, dermatophilosis (streoto-thricosis), east coast fever and other tick-borne diseases, and foot and mouth disease. Brucellosis occurs widely and parasitic gastro-enteritis is common, and takes a heavy toll of calves under low management level. Fairly satisfactory control measures for a number of these diseases are available but continued vigilance is necessary to ensure that herds receive protection.

Foot and mouth disease is not important at a low level of production although its occurrence may prevent the export of meat. Ticks can be controlled by dipping or spraying but the provision of facilities and supervision is sometimes difficult.

Sheep and goats are susceptible to a variety of diseases including bacterial pneumonia, internal parasites, foot-rot and in the case of goats caprine pleuropneumonia and in sheep, sheep pox. Treatment is not normally available or sought and losses can be heavy although sick animals are killed and the carcase utilized.

Camels are very susceptible to tick-borne disease and trypanosomiasis. However, they are rarely kept in zones with more than 90 days growing period.

The distribution of trypanosomiasis and its tse-tse vector in Kenya has been mapped and is included in the land resources data base (Chapter 3). It has been assumed that in the thermal zones 1, 2, 3 and 4, loss in livestock production performance would be of the order of 75% in the low technology systems and 50% in intermediate and high technology systems due to trypanosomiasis.

6.4 Estimation of Feed Requirements

Part IV of the livestock productivity model (Figure 6.1) formulates the livestock feed requirements, taking into account maintenance and production needs.

Feed requirements have been formulated to support the herd performances quantified in Section 6.3 for the individual livestock systems.

In order to support the body's processes and promote production, animals must consume regular supplies of various nutrients. These nutrients may be broadly defined as energy (from carbohydrates and fats), protein, vitamins, minerals and water. They are contained in animal feeds, which are largely of plant origin, in different concentration and combination. Under most intensive systems of animal husbandary, the animal may not always be able to obtain a balanced diet throughout the year because of the seasonal variation in the composition of the herbage.

Water is also needed by the animal, this is obtained from three sources: (a) drunk as water, (b) contained in the herbage or other feed, and (c) resulting from the oxidation of carbohydrates in the tissues. Availability of water is a problem in some parts of Kenya, and much of the pastoral zone has limited permanent water forcing nomadic behaviour. Certain stretches of the country have no water resources. Such a situation must be taken into account in final estimates of livestock carrying capacities. The available sources of information include data which would enable these areas to be identified and measured at the district level and this should be incorporated in the model for a more effective treatment.

TABLE 6.17
Feed requirements per herd TLU (kg/day dry weight)

Livestock system	Inputs level		
	Low	Intermediate	High
Pastoral (<120 days LGP)			
- Cattle	7.0	7.2	7.4
- Goat	6.6	6.8	7.0
- Sheep	6.6	6.8	7.0
- Camel	6.5	6.6	6.7
Non-pastoral (>120 days LGP)			
- Cattle	7.8	8.5	8.9[1]
- Goat	10.0	11.5	16.1[2]
- Sheep	9.1	11.3	11.6

1 Includes 1.2 kg/day primary products (3.2 kg/day per lactating cow)
2 Includes 4.8 kg/day primary products (0.6 kg/day per lactating doe).

A summary of reference feed requirements per herd TLU is given in Table 6.17 for non-pastoral and pastoral systems for three levels of inputs situations. In the non-pastoral systems, intake requirements for cattle, goat and sheep are based on field verification for the herd structures presented in Section 6.2 for the performance output levels described in Section 6.3. In the model, crop residue intake in the non-pastoral systems is limited to 30%, 20%, and 10% of total feed intake respectively in the low, intermediate and high technology system.

For pastoral systems, feed requirements are based on Boudet and Riviere (1968) for the herd structures and performances presented in Sections 6.2 and 6.3 For poultry and pig, the standard requirements are used (FAO 1988b).

Feed requirements for each system are presented hereunder.

6.4.1 Cattle Systems: Dairy and Meat

In the low technology system, one cow unit requires about 3,740 kg dry matter (DM), corresponding to 1,650 kg total digestable nutrients (TDN) and 210 kg digestible crude protein (DCP) per year, for maintenance and production. These feed requirements are met by grazing Kikuyu/star grass pasture and maize stover.

In the intermediate technology system, one cow unit requires about 5,200 kg DM (2,560 kg TDN and 300 kg DCP per year) for maintenance and production. These feed requirements are met by grazing Kikuyu/star grass pasture and maize stover.

In the high technology system, one cow unit requires about 7,200 kg DM (3500 kg TDN and 420 kg DCP per year) for maintenance and production. This is provided by the napier/bana grass, by feeding maize stover and by feeding 1,165 kg concentrates.

The above requirements correspond to 7.8, 8.5 and 8.9 kg/day per reference herd TLU for the low, intermediate and high technology systems respectively (Table 6.17).

6.4.2 Goat Systems: Dairy and Meat

In the low technology system, one doe unit requires about 470 kg DM per year, provided by natural pasture.

In the intermediate technology system, one doe unit requires about 700 kg DM per year. This is provided by a combination of sources: natural pasture (280 kg), fodder crops (280 kg) and crop residue (140 kg).

In the high technology system, one doe unit requires about 960 kg DM per year. This is provided by fodder crops (610 kg), crop residue (140 kg) and concentrates (210 kg).

The above requirements correspond to 10.0, 11.5 and 16.1 kg/year per reference herd TLU for the low, intermediate and high technology systems respectively (Table 6.17).

6.4.3 Sheep Systems: Meat and Wool

In the low technology system, one ewe unit requires about 360 kg DM per year, provided by natural pasture.

In the intermediate technology system, one ewe unit requires about 610 kg DM per year. This is provided by a combination of sources: natural pasture, fodder crops and crop residue.

In the high technology system, one ewe unit requires about 750 kg DM per year. This is provided by natural pasture, fodder crops, crop residue and concentrates.

The above requirements correspond to 9.1, 11.3 and 11.6 kg/year per reference herd TLU for the low, intermediate and high technology systems respectively (Table 6.17).

6.4.4 Pastoral Systems: Milk and Meat

For the pastoral systems (< 120 days growing period), feed requirements are based on a daily intake of 2.5 kg dry matter per 100 kg liveweight or 6.25 kg dry matter for the 250 kg reference TLU. Maintenance requirements are 2.9 FU/day and 160 g/day digestible protein (DP). The annual maintenance dietary needs of a reference TLU are thus 1,060 FU or 2,280 kg DM (1FU = 2.15 kg DM) and 58 kg DP. Production requirements are in addition at 350 extra FU/year (0.95 FU/day) and 28 kg DP (75 g/day) for weight gain of 100 kg/year (300 g/day) or a production of 1,000 kg/year(2.74 kg/day) of milk.

The above requirements correspond to 7.0, 7.2 and 7.4 kg DM/day per TLU for the low, intermediate and high technology systems respectively for cattle; 6.6, 6.8 and 7.0 kg DM/day per TLU for goat and sheep; and 6.5, 6.6 and 6.7 kg DM/day per TLU for camel (Table 6.17).

6.4.5 Poultry and Pig: Meat and Egg

These animals are considered only under the intensive systems and standard requirements are used (FAO 1988b). For poultry these are 2.5 kg of feed (primary products) for 1 kg of meat, and 3,5 kg of feed for 1 kg of egg mass. For pig it is 4 kg of feed for 1 kg of meat.

6.5 Livestock Productivity Potential

Part V of the livestock productivity model (Figure 6.1) deals with quantification of livestock productivity potential (secondary productivity) of land (agro-ecological cells). This is achieved by setting feed requirements of livestock systems from Part IV against feed supply from Part I.

However, before it is possible to set feed requirements against feed supply, the latter from its various sources as applicable must be quantified by agro-ecological cell in relation to the objective function driving the model.

The permissible thermal and LGP zones for the different livestock systems is taken from the Tables 6.14 and 6.15, and the expected output of the products per herd TLU is taken from Table 6.16 for cattle, goat, sheep and camel, and from Section 6.4.5 for poultry and pig. Where output performance is assumed to be affected by constraints such as temperature stress, tse-tse, the expected loss in performance output is taken into account.

Chapter 7

Fuelwood productivity

This chapter describes the fuelwood productivity model. The model is schematically shown in Figure 7.1, and is structured along the lines of the crop suitability model[1] of the FAO Agro-ecological Zones Project (FAO 1978-81). It is applied within the framework of land evaluation guidelines for forestry (FAO 1984c). It comprises of the following activities:

(i) Selection of tree species and definition of land utilization types (LUTs) (e.g. species; technology and input level; labour; capital; markets).

(ii) Determination of climatic requirements of species and LUTs and matching climatic requirements with the characteristics of the inventoried climatic zones (thermal zones and growing period zones), and quantifying the climatically attainable yield potentials.

(iii) Determination of edaphic (soil) requirements of species and LUTs, and matching edaphic requirements with the characteristics of the inventoried soil units, textures, phases and stoniness to rate edaphic limitations.

(iv) Quantifying soil erosion hazards (topsoil loss) in each climate-soil unit of the land resources inventory by LUT and the associated productivity losses.

(v) Modifying the climatic yield potentials (in ii) according to soil limitations (in iii) and erosion hazards (iv) to quantify yield potentials and ecological land suitabilities of each inventoried climatic-soil land unit for each LUT.

The model operates on the land resources database described in Chapter 3. Each of the above activities are described in the following sections.

7.1 Tree Species and Land Utilization Types

A total of 31 species are included in the model. They are listed in Table 7.1 together, with information on height, coppicing ability, nitrogen fixing ability, density, calorific value and utilization.

1 In the AEZ rainfed suitability model, the term productivity normally refers to land's production potential for the total length of growing period over years. Perennial fuelwood species utilize all the time available in a growing period over the rotation age so that the term fuelwood productivity is synonymous with the term fuelwood yield.

FIGURE 7.1
Schematic presentation of fuelwood productivity model

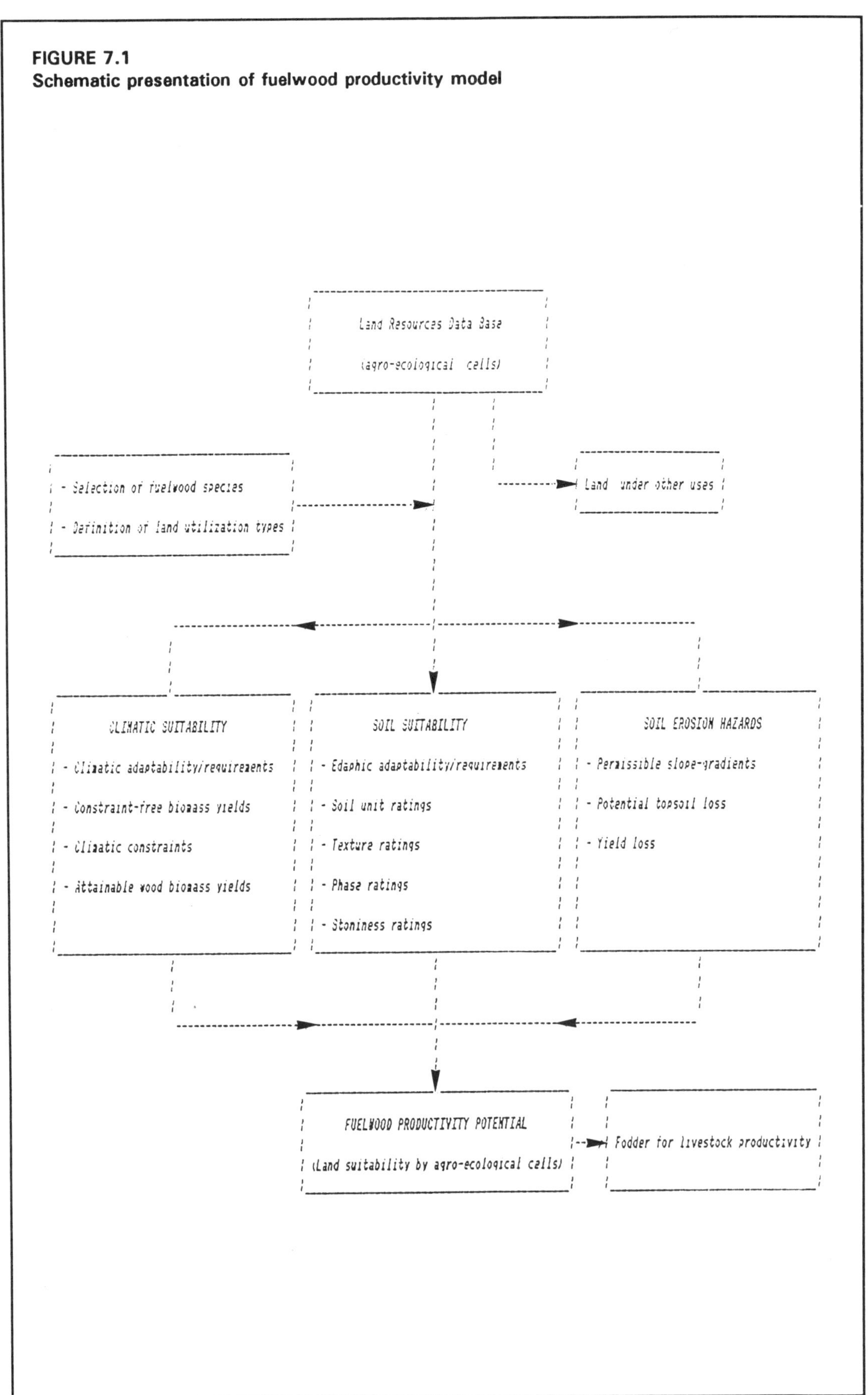

TABLE 7.1
Fuelwood species characteristics

Species	Mature height (m)	Coppicing ability	N-fixer	Density (gm/cm^3)	Calorific value (Kcal/kg)	Utilization
Acacia albida	20-30		✓			Fo
Acacia gerradii	10-15		✓			Fo
Acacia nilotica	15-20	+ +	✓	0.65-0.70	4800-4950	C,Fo,G,Ho,S
Acacia senegal	2-5	+ +	✓			C,Fo,G,T
Acacia tortilis	4-10		✓			Fo
Bridelia micrantha	8-15					C,T
Calliandra calothyrsus	8-12	+ +	✓	0.50-0.80	4500	Fo,Ho,M,Or
Calodendrum capense	10-20					C,Or,T
Cassia siamea	15-20	+ +		0.60-0.80	4500-4600	C,D,Fo*,H,M,Or,S,T,Wb
Casuarina equisetifolia	25-30	+	✓	0.80-1.20	4950	C,D,Fo,P,T,Wb
Casuarina cunninghamiana	25-30		✓			C
Conocarpus lancifolius	15-18		✓	0.81		C,Fo,T
Croton megalocarpus	15-25		✓			C,T
Cupressus lucitanica	25-30			0.45-0.48		–
Eucalyptus camaldulensis	30-40	+ +		0.55-0.85	4800	C,Fo,H,Ho,Or,P,S,T,Wb
Eucalyptus citriodora	30-40	+		0.65-1.10	4750	C,Fo,Ho,O.P.T
Eucalyptus globulus	40-50	+ +		0.55-0.78		C,P,Pl,T
Eucalyptus grandis	40-55	+ +		0.48-0.64		C,P,Pl,T
Eucalyptus microcorys	25-30	+ +		0.90-0.99		T
Eucalyptus microtheca	10-20	+		0.75-0.85		C,T
Eucalyptus saligna	35-45	+ +		0.48-0.64		C,O,P,Pl,T
Eucalyptus tereticornis	35-45	+ +		0.65-1.05	4800	C,Fo,Ho,O,P,Pl,S,Sb,T
Faurea saligna	5-10					C,Ho,T
Gliricidia sepium	5-15	+ +	✓	0.40-0.65	4900	C,Fb,Fo*,Ho,M,Or,S,T
Grevillea robusta	25-35		✓	0.54-0.66		C,Pl,T
Leucaena leucocephala	10-20	+ +	✓	0.50-0.70	4200-4600	C,Fo*,M,Or,P,S,T
Oleo africana	3-6					C,Fr,T
Parkinsonia aculeata	4-5	+ +				Fo,Wb
Prunus africanum	25-30					T
Sesbania sesban	4-8	+ +	✓	0.40-0.50	4500-4600	Fo
Tamarindus indica	20-30		✓			C,T

Sources: C.M. Ndegwa, Project GCP/KEN/051/AUL, 'Fuelwood afforestation and extension in Baringo', pers. comm. 1988; Davidson 1985; Teel 1984; Skerman 1982; Goehl 1981; Webb, Wood and Smith 1980; FAO 1981.

Coppicing ability: + + = good; + = fair; no entry indicates poor or none.

Utilization: C = charcoal; D = dye; Fb = firebreak; Fo = fodder; Fo* = fodder (but potentially toxic); Fr = fruit; G = gum; H = hedge; Ho = honey; M = green manure; O = oil; Or = ornamental; P = pulp (wood); Pl = plywood, board, etc.; S = shade; Sb = shelterbelt; T = timber, etc.; Wb = windbreak.

Each tree species is considered for fuelwood production at three levels of inputs circumstances (low,intermediate and high). The attributes of the three input level production circumstances are listed in Table 7.2, and they form the basis of the definition of the land utilization types.

7.2 Climate Adaptability and Yield Potential

Understanding the relationships between the climatic environment and ecophysiological processes of growth, development and yield in trees forms the basis of formulating quantitative descriptions of the climatic adaptability of improved and unimproved provenances and their productivity potentials in land use. Principles of climatic adaptability for plants are described in Kassam, Kowal and Sarraf (1977) and (FAO 1978-81).

Photosynthesis produces the source of assimilates which plants use for growth, and the rate of photosynthesis is influenced by both temperature and radiation. However, plants are

TABLE 7.2
Attributes of LUTs considered for fuelwood production

Attribute	Low inputs	Intermediate inputs	High inputs
Produce and production	Rainfed production of fuelwood tree species for firewood or charcoal		
Market orientation	Subsistence production	Subsistence production plus commercial sale of surplus	Commercial production
Capital intensity	Low	Intermediate, with credit on accessible terms	High
Labour intensity	High, including uncosted family labour	Medium, including uncosted family labour	Low, family labour costed if used
Power source	Manual labour with hand tools	Manual labour with handtools, some mechanization	Complete mechanization, including harvesting
Technology	Local provenancès; no agro-chemicals; minimum conservation measures	Improved provenances as available; appropriate extension packages, including some fertilizer application and pest and disease control; some conservation measures	High yielding provenances; optimum fertilizer use; chemical pest and disease control; full conservation measures
Infrastructure	Market access not necessary; inadequate advisory services	Some market accessibility necessary, with access to nurseries, demonstration plots and advisory services	Market access essential; high level of nursery and advisory services; application of research findings
Land holding	Small, fragmented	Small, sometimes fragmented	Large, consolidated
Income level	Low	Moderate	High

also obliged to undergo sequences of phenological and morphological developments in time and space to allow photosynthetic assimilates to be converted into growth of plant parts and economically usefull yields of satisfactory quantity and quality. The development sequence of tree growth in relation to the calendar (i.e. tree phenology) is influenced by climatic factors.

In general, temperature determines the rate of growth and development of plant parts and the tree as a whole. However, in some tree species, temperature may also determine whether a particular development process will begin or not (e.g., chilling requirement for bud formation and floral initiation), the time when bud break will occur, the subsequent rate of development and the time when the process will stop (Cannell and Last 1976).

In the seasonally dry climates of Kenya, ability to survive the dry period is an important adaptability characteristic just as frost hardiness is for survival in the cooler thermal zones at higher altitudes.

Accordingly, in assessments of land suitabilities, consideration has to be given to the specific climatic requirements and adaptability for survival, growth and development.

7.2.1 Photosynthesis Characteristics

All the fuelwood species listed in Table 7.1 have C3 photosynthesis pathway and are classified into two adaptability groups (Table 7.3). Group I species are adapted to operate in

TABLE 7.3
Adaptability groups for fuelwood species

Characteristics	Group I (<20°C)		Group II (>20°)	
Photosynthetic pathway	C_3		C_3	
Rate of maximum photosynthesis (P_{max}) (kg CH_2O/ha/hr)	5-30		5-30	
Optimum temperature (mean) for P_{max} (°C)	15-20		20-30	
Productivity class A (P_{max} = 5-10 kg CH_2O/ha/hr)	*Acacia gerrardii*	(N)	*Acacia albida*	(N)
	Croton megalocarpus	(N)	*Acacia nilotica*	(N)
	Grevillea robusta	(N)	*Acacia tortilis*	(N)
	Oleo africana		*Calliandra calothyrus*	(N)
			Conocarpus lancifolius	(N)
			Gliricidia sepium	(N)
			Tamarindus indica	(N)
Productivity class B (P_{max} = 10-20 kg CH_2O/ha/hr)	*Bridella micrantha*		*Bridella micrantha*	
	Calodendrum capense		*Cassia siamea*	
	Casuarina cunninghamiana	(N)	*Casuarina equisetifolia*	(N)
	Cupressus lucitanica		*Eucalyptus citriodora*	
	Eucalyptus microcorys		*Eucalyptus microtheca*	
	Faurea saligna		*Eucalyptus tereticornis*	
	Prunus africanum		*Parkinsonia aculeata*	
Productivity class C (P_{max} = 20-30 kg CH_2O/ha/hr)	*Eucalyptus globulus*		*Eucalyptus camaldulensis*	
	Eucalyptus saligna		*Eucalyptus grandis*	
	Sesbania sesban	(N)	*Eucalyptus saligna*	
			Leucaena leucocephala	(N)
			Sesbania sesban	(N)

(N) - Nitrogen fixer

cooler conditions (mean temperatures 10-20 °C), whereas Group II species are adapted to operate in warmer conditions (mean temperatures 20-30 °C). Both groups have species with nitrogen fixing capability.

Rates of maximum photosynthesis (Pm) for both adaptability groups are in the range 5-30 kg CH_2O ha^{-1} hr^{-1} (Landsberg 1986). Species in each adaptability group are therefore further classified into three photosynthesis productivity classes. They are class A, Pm = 5-10 kg CH_2O ha^{-1} hr^{-1}; class B, Pm = 10-20 kg CH_2O ha $^{-1}$ hr^{-1} : and class C, Pm = 20-30 kg CH_2O ha^{-1} hr^{-1}. These photosynthesis rates of productivity class A, B, and C correspond to mean annual total (including foliage, stem and roots) biomass increments of 12.5-25.0, 25.0-40.0 and 40.0-60,0 t/ha dry weight respectively or annual wood biomass (stem and branch wood) increments of 8.0-15.0, 15.0-25.0 and 25.0-40.0 t/ha dry weight respectively. The relationships between photosynthesis and temperature for these six adaptability classes are presented in Table 7.4.

7.2.2 Rotation Length

Rotation length in the model is taken as the age at 'maximum yield ', and is the point when annual increment is equal to mean annual increment over the total period since establishment (Nilsson 1983).

TABLE 7.4
Relationships between temperature and rate of photosynthesis (kg CH_2O/ha/hr) for six adaptability classes of fuelwood species

Adaptability class	Temperature (°C)							
	5	10	15	20	25	30	35	40
I - A	0.75	3.0	6.0	7.5	7.5	6.0	3.0	1.5
I - B	1.5	6.0	12.0	15.0	15.0	12.0	6.0	3.0
I - C	2.5	10.0	20.0	25.0	25.0	20.0	10.0	5.0
II - A	-	0.75	4.0	6.0	6.0	7.5	6.0	4.0
II - B	-	1.5	8.0	12.0	12.0	25.0	12.0	8.0
II - C	-	2.5	15.0	20.0	20.0	25.0	20.0	15.0

TABLE 7.5
Rotation length (years) by moisture zones

Photosynthetic productivity class	Semi-arid 60-119 days	Dry Sub-humid 120-179 days	Moist Sub-humid 180-269 days	Humid ≥270 days
A	15.0-17.5	12.5-15.0	10.0-12.5	7.5-10.0
B	12.5-15.0	10.0-12.5	7.5-10.0	5.0-7.5
C	10.0-12.5	7.5-10.0	5.0-7.5	< 5.0

Rotation length is affected by the photosynthesis productivity class of the species and by length of growing period. Rotation lengths applied in the model are given in Table 7.5.

7.2.3 Climatic Yield Potentials

Thermal zone ratings for each of the species are given in Table 7.6. Five suitability classes are employed (i.e., S1, very suitable; S2, suitable; S3, moderately suitable; S4, marginally suitable; and N, not suitable), and the ratings apply to production at all the three levels of inputs.

A rating of S1 indicates that the temperature conditions for tree growth and development are optimal, and that it is possible to achieve the maximum attainable silvicultural yield potential provided there are no moisture or soil-landform limitations. A rating of S2 indicates that there are moderate temperature constraints to growth and development, and that there would be a suppression of yield potential of the order of 25%. A rating of S2 indicates that there are moderate to severe temperature constraints, and that there would be a yield suppression of the order of 50%. A rating of S4 indicates that yield suppression of the order of 75%. A rating of N indicates that temperature conditions are not suitable for production.

Growing period zones which have been considered for yield assessments for each species are shown in Table 7.7 which represents a moisture screen. Table 7.7 is based on the actual research information obtained from local experiments and permanent sample plots.

Ecophysiological models have not been widely applied in the estimation of stand and site productivity potentials. A useful description of the state-of-the art is given in Landsberg (1986). Potential attainable yields (total and wood biomass) were derived according to the

TABLE 7.6
Thermal zone suitability ratings for fuelwood species

Species	T1 >25°	T2 22.5-25.0°	T3 20.0-22.5°	T4 17.5-20.0°	T5 15.0-17.5°	T6 12.5-15.0°	T7 10.0-12.5°	T8 5.0-10.0°	T9 <5.0°
Acacia albida	S1	S1	S1	S3	S4	N	N	N	N
Acacia gerradii	S4	S3	S1	S1	S1	S1	S3	N	N
Acacia nilotica	S1	S1	S1	S1	S3	N	N	N	N
Acacia senegal	S1	S1	S1	S3	S4	N	N	N	N
Acacia tortilis	S1	S1	S1	S3	S4	N	N	N	N
Bridelia micrantha	S1	S1	S1	S1	S1	S1	S3	N	N
Calliandra calothyrsus	S1	S1	S1	S3	S4	N	N	N	N
Calodendrum capense	S4	S3	S1	S1	S1	S1	S3	N	N
Cassia siamea	S1	S1	S1	S3	S4	N	N	N	N
Casuarina equisetifolia	S1	S1	S1	S1	S3	N	N	N	N
Casuarina cunninghamiana	S4	S3	S1	S1	S1	S1	S3	N	N
Conocarpus lancifolius	S1	S3	S4	N	N	N	N	N	N
Croton megalocarpus	S4	S3	S1	S1	S1	S1	S3	N	N
Cupressus lucitanica	N	S4	S3	S1	S1	S1	S3	N	N
Eucalyptus camaldulensis	S1	S1	S1	S3	N	N	N	N	N
Eucalyptus citriodora	S1	S1	S1	S1	S3	N	N	N	N
Eucalyptus globulus	S4	S3	S1	S1	S1	S3	N	N	N
Eucalyptus grandis	S1	S1	S1	S1	S3	N	N	N	N
Eucalyptus microcorys	S4	S3	S1	S1	S1	S1	S3	N	N
Eucalyptus microtheca	S1	S1	S1	S3	N	N	N	N	N
Eucalyptus saligna	S1	S1	S1	S1	S1	S1	S3	N	N
Eucalyptus tereticornis	S1	S1	S1	S1	S3	N	N	N	N
Faurea saligna	S4	S3	S1	S1	S1	S1	S2	N	N
Gliricidia sepium	S1	S1	S1	S1	S3	N	N	N	N
Grevillea robusta	S4	S3	S1	S1	S1	S1	S2	N	N
Leucaena leucocephala	S1	S1	S1	S3	S4	N	N	N	N
Oleo africana	N	S4	S3	S1	S1	S1	S3	N	N
Parkinsonia aculeata	S1	S1	S1	S3	S4	N	N	N	N
Prunus africanum	N	S4	S1	S1	S1	S1	S3	N	N
Sesbania sesban	S1	S1	S1	S1	S1	S1	S3	N	N
Tamarindus indica	S1	S1	S1	S3	S4	N	N	N	N

method developed by the FAO-AEZ Project (Kassam 1977; FAO 1978-81), and modified to take into account the generally accepted fact that for fuelwood tree species, total biomass yield at 50% of rotation length is 38% of the standing total biomass yield at 100% rotation length (Nilsson 1983).

It is assumed that wood biomass (stem wood and branch wood) is 0.6 of total biomass, foliage biomass 0.2 and root biomass 0.2. Partitioning of total wood biomass into main stem and branch wood biomass is assumed to be in the ratio of 0.8 and 0.2. Leaf area index at maximum annual growth rate is assumed to be 5 or more, and the period of annual growth is equal to the inventoried lengths of growing period. These reference model variables can be modified as appropriate to take into account differences between species and environmental conditions.

Total biomass productivity estimates (Bm) in terms of mean annual increments (t/ha dry weight) are given in the Appendix in Table A7.1 for high level of inputs by length of growing period for species with and without nitrogen fixing ability for the three photosynthesis productivity classes. For the low level of inputs circumstance, site yield potentials are assumed to be 50% of those at the high level. At intermediate level of inputs, yield potentials are assumed to be half-way between the low and the high levels of inputs.

TABLE 7.7
Moisture screen for fuelwood species

Species	Length of Growing Period (LGP) (Days)														
	0	1-29	30-59	60-89	90-119	120-149	150-179	180-209	210-239	240-269	270-299	300-329	330-364	365⁻	365⁺
Acacia albida			●	●	●	●	●	●	●						
Acacia gerradii					●	●	●	●	●	●	●				
Acacia nilotica			●	●	●	●	●	●	●	●					
Acacia senegal			●	●	●	●	●	●	●						
Acacia tortilis			●	●	●	●	●	●	●	●					
Bridelia micrantha						●	●	●	●	●	●	●	●	●	●
Calliandra calothyrsus								●	●	●	●	●	●	●	●
Calodendrum capense							●	●	●	●	●	●	●	●	●
Cassia siamea					●	●	●	●	●	●	●				
Casuarina equisetifolia					●	●	●	●	●	●	●				
Casuarina cunninghamiana						●	●	●	●	●	●	●	●	●	●
Conocarpus lancifolius			●	●	●	●	●	●	●	●					
Croton megalocarpus						●	●	●	●	●	●				
Cupressus lucitanica								●	●	●	●	●			
Eucalyptus camaldulensis					●	●	●	●	●	●					
Eucalyptus citriodora						●	●	●	●	●	●				
Eucalyptus globulus							●	●	●	●	●	●			
Eucalyptus grandis								●	●	●	●	●	●	●	●
Eucalyptus microcorys							●	●	●	●	●				
Eucalyptus microtheca			●	●	●	●	●	●	●	●					
Eucalyptus saligna							●	●	●	●	●	●	●	●	●
Eucalyptus tereticornis					●	●	●	●							
Faurea saligna						●	●	●	●	●	●	●	●	●	●
Gliricidia sepium							●	●	●	●	●	●	●	●	●
Grevillea robusta						●	●	●	●	●	●	●	●	●	●
Leucaena leucocephala						●	●	●	●	●	●	●	●	●	●
Oleo africana						●	●	●	●	●	●				
Parkinsonia aculeata			●	●	●	●	●								
Prunus africanum							●	●	●	●	●	●	●	●	●
Sesbania sesban						●	●	●	●	●	●	●	●	●	●
Tamarindus indica			●	●	●	●	●	●	●	●					

Total biomass productivity for intermediate and low levels of inputs are given in the Appendix in Tables A7.2 and A7.3 respectively.

Wood biomass yield estimates (Bw) in terms of mean annual increments (t/ha dry weight) are given in the Appendix in Tables A7.4, A7.5 and A7.6 respectively for high, intermediate and low levels inputs circumstances. Wood biomass estimates in Tables A7.4, A7.5 and A7.6 apply in the growing period zones allowable by the moisture screen in Table 7.7.

All tree species are matched to total lengths of L1, L2, L3 and L4. Yields in Tables A7.1 to A7.6 apply to years with normal length of growing period, i.e. growing period with a humid period during which precipitation is greater that potential evapotranspiration. For years with intermediate growing periods, i.e. growing periods with no humid period, full water requirements cannot be met and yield reductions are assumed to be of the order of 50% on all soils except Fluvisols and Gleysols. The percentage of occurrence of intermediate lengths of growing periods in all LGP-Pattern zones is 100% in LGP zone 1-29 days; 65% in LGP zone 30-59 days; 25% in LGP zone 60-89 days; 10% in LGP zone 90-119 days; and 5% in LGP zone 120-149 days.

At this stage in the model development, it has not been possible to take into account in the climatic suitability assessment other climatically driven constraints such as pest and diseases and workability, which may reduce yield. It should be possible to take such constraints into account in the future as the information and research base for fuelwood production improves.

An exception to the general methodology for climatic suitability assessment applies to areas occupied by Fluvisols because the length of growing period does not fully reflect their particular circumstance with regards to moisture regime. Fluvisols ratings are presented in Technical Annex 6 for the three levels of inputs circumstances.

7.3 Edaphic Adaptability and Suitability

In order to assess soil suitability for fuelwood production, the soil requirements of tree species must be determined. Further, these requirements must be understood within the context of limitation imposed by landform and other features (e.g. soil phases) which do not form part of soil composition but have a significant influence on the use that can be of the soil.

7.3.1 Basic Soil Requirements

Basic soil requirements for fuelwood tree species relate to soil properties (described in Section 5.1.3.2).

From the basic soil requirements for forestry land use, a number of responses related to soil characteristics have been derived for the fuelwood species.

The correlation between the basic soil requirements listed in Section 5.1.3.2 and soil characteristics given in Table 5.9 has been used as soil factors to rate tree crop performance and soil requirements for fuelwood species are summarized in Table 7.8.

TABLE 7.8
Soil requirements of fuelwood species

Species	Texture		Drainage		Soil depth (cm)		Reaction (pH)		Salinity (mmhos)		Flooding	
	optimum	range	optimum	range	optimum	marginal	optimum	range	optimum	marginal	optimum	marginal
Acacia albida	SL-SC	LS-KS	W	MW-SE	>120	75-120	5.5-7.0	5.0-7.5	<4	4-6	FO	F1
Acacia gerradii	L-C	SL-KC	MW-W	I-SE	>120	75-120	5.5-7.0	5.0-7.5	<4	4-6	F1	F2
Acacia nilotica	L-C	SL-KC	W	MW-SE	>120	75-120	5.5-7.5	5.0-8.0	<4	4-8	F1	F2
Acacia senegal	SL-SC	LS-KC	W	MW-SE	>120	75-120	5.5-7.0	5.0-7.5	<4	4-6	FO	F1
Acacia tortilis	SL-SC	LS-KC	W	MW-SE	>75	50-75	6.5-8.0	6.0-8.5	<4	4-8	FO	F1
Bridelia micrantha	SL-SC	LS-KC	W	MW-SE	>120	75-120	5.5-7.0	5.0-7.5	<4	4-6	FO	F1
Calliandra calothyrsus	SL-SC	LS-KC	W	MW-SE	>120	75-120	6.0-7.0	5.5-7.5	<4	4-6	FO	F1
Calodendrum capense	SL-SC	LS-KC	W	MW-SE	>120	75-120	5.5-7.0	5.0-7.5	<4	4-6	FO	F1
Cassia siamea	SL-CL	LS-SiC	W	MW-SE	>150	100-150	6.0-7.0	5.5-7.5	<4	4-6	FO	F1
Casuarina equisetifolia	SL-SiL	LS-CL	W	MW-SE	>120	75-120	6.5-8.0	6.0-8.5	<12	12-16	F1	F2
Casuarina cunninghamiana	SL-SC	LS-KC	W	MW-SE	>120	75-120	5.5-7.0	5.0-7.5	<4	4-6	F1	F2
Conocarpus lancifolius	SL-CL	LS-SC	MW-W	I-SE	>120	75-120	6.5-8.0	6.0-8.5	<8	8-12	F1	F2
Croton megalocarpus	SL-SC	LS-KC	W	MW-SE	>120	75-120	6.5-8.0	6.0-8.5	<4	4-8	FO	F1
Cupressus lucitanica	L-CL	SL-SC	W	MW-SE	>150	100-150	5.5-7.0	5.0-7.5	<4	4-6	FO	F1
Eucalyptus camaldulensis	SL-SC	LS-KC	MW-W	I-SE	>120	75-120	5.0-6.5	4.5-7.5	<2	2-4	F1	F2
Eucalyptus citriodora	SL-CL	LS-SiC	W	MW-SE	>150	100-150	5.5-7.0	5.0-7.5	<4	4-6	FO	F1
Eucalyptus globulus	L-C	SL-KC	W	MW-SE	>150	100-150	5.5-7.0	5.0-7.5	<2	2-4	FO	F1
Eucalyptus grandis	SL-CL	LS-SiC	W	MW-SE	>120	75-120	5.5-7.0	5.0-7.5	<4	4-6	FO	F1
Eucalyptus microcorys	L-CL	SL-SC	W	MW-SE	>120	75-120	5.5-7.5	5.0-8.0	<4	4-8	F1	F2
Eucalyptus microtheca	L-C	SL-KC	MW-W	I-SE	>120	75-120	6.5-8.0	6.0-8.5	<4	4-8	F1	F2
Eucalyptus saligna	SL-CL	LS-SiC	W	MW-SE	>120	75-120	5.5-7.0	5.0-7.5	<4	4-6	FO	F1
Eucalyptus tereticornis	SL-SC	LS-KC	W	MW-SE	>120	75-120	5.5-7.0	5.0-7.5	<4	4-6	FO	F1
Faurea saligna	SL-SC	LS-KC	W	MW-SE	>120	75-120	5.5-7.0	5.0-7.5	<4	4-6	FO	F1
Gliricidia sepium	SL-SC	LS-KC	W	MW-SE	>120	75-120	5.0-6.5	4.5-7.0	<2	2-4	FO	F1
Grevillea robusta	SL-CL	LS-SiC	W	MW-SE	>120	75-120	5.5-7.0	5.0-7.5	<4	4-6	FO	F1
Leucaena leucocephala	SL-SC	LS-KC	W	MW-SE	>75	50-75	5.5-7.0	5.5-8.5	<8	8-12	FO	FO
Oleo africana	SL-SC	LS-KC	W	MW-SE	>120	75-120	6.5-8.0	6.0-8.5	<4	4-8	FO	F1
Parkinsonia aculeata	SL-CL	LS-SiC	W	MW-SE	>120	75-120	6.5-8.0	6.0-8.5	<8	8-12	FO	F1
Prunus africanum	SL-SC	LS-KC	W	MW-SE	>120	75-120	5.5-7.0	5.0-7.5	<4	4-6	FO	F1
Sesbania sesban	SL-SC	LS-KC	MW-W	I-SE	>120	75-120	6.0-7.0	5.5-7.5	<12	12-16	F1	F2
Tamarindus indica	SL-SC	LS-KC	W	MW-SE	>120	75-120	5.5-7.5	5.0-7.5	<4	4-6	FO	F1

Texture classes: LS = loamy sand; SL = sandy loam; SiL = silty loam; L = loam; CL = clay loam; SC = sandy clay; SiC = silty clay; C = clay; KC = kaolinitic clay.
Drainage classes: I = imperfectly; MW = moderately well; W = well; SE = somewhat excessive.
Flooding classes: FO = no flooding; F1 = occasional floods; F2 = frequent floods.

As exlained earlier the soil units (Table 3.16) have been defined in terms of measurable and observable properties of the soil itself; and specific clusters of such properties are combined into 'diagnostic horizons' and 'diagnostic properties'. They are also used in rating soil suitability.

7.3.2 Edaphic Suitability

The edaphic suitability classification is input-specific and based on:

(i) matching the soil requirements of fuelwood species with the soil conditions of the soil units described in the soil inventory (soil unit evaluation); and

(ii) modification of the soil unit evaluation by limitations imposed by texture, phase and slope conditions.

The soil unit evaluation for fuelwood species production is expressed in terms of ratings based on how far the soil conditions of a soil unit meet the growth and production requirements under a specified level of inputs. The appraisal is effected in five basic classes for each species and level of inputs, i.e. very suitable (Sl), suitable (S2), moderately suitable (S3), marginally suitable (S4), and not suitable (N).

A rating of Sl indicates that the soil conditions are optimal, and that suppression of potential yields (if any) are assumed to be nil or slight. A rating of S2 indicates that there are slight to moderate soil constraints and there would be a suppression of potential yields of the order of 25%. A rating of S3 indicates that there are moderate to severe soil constraints and there would a suppression of potential yields of the oder of 50%. A rating of S4 indicates that there are severe soil constraints and there would be a suppression of potential yields of the order of 75%. A rating of N indicates that soil conditions are not suitable for production.

The soil unit ratings for fuelwood are given in Technical Annex 6, and apply as given, provided there are no additional limitations imposed by soil texture, phase and stoniness. Modifications are required where such limitations are present.

In the case of soil texture, soil unit ratings remain the unchanged if the soil is an Albic, Cambic, Ferralic, Calcaro-Cambic or Luvic Arenosol (Q, Qa, Qc, Qf, Qkc, Ql) or a vitric Andosol (Tv), or where textures are medium (fine sandy loam, sandy loam, loam, sandy clay loam, clay loam, silty clay loam, silt), or fine (sandy clay, silty clay, peaty clay, clay). In all other cases (i.e. with coarse textures: sand, loamy coarse sand, fine sand, loamy fine sand, loamy sand) the soil unit rating is one class (25%) lower.

Limitations imposed by phase and stoniness are rated using the five basic classes already described. The ratings are presented in Technical Annex 6.

7.4 Slope Limitations and Soil Erosion

Limitations imposed by slope are taken into account in three steps (Chapter 4). Step one defines the slopes which are permissible for fuelwood production, and as a model variable this is defined as slopes less than 45% (Table 4.1).

Step two involves the computation of potential topsoil loss which is estimated, by input level, through a modified Universal Soil Loss Equation (Wischmeier and Smith 1978).

FIGURE 7.2
Schematic presentation of the land suitability assessment programme for fuelwood production

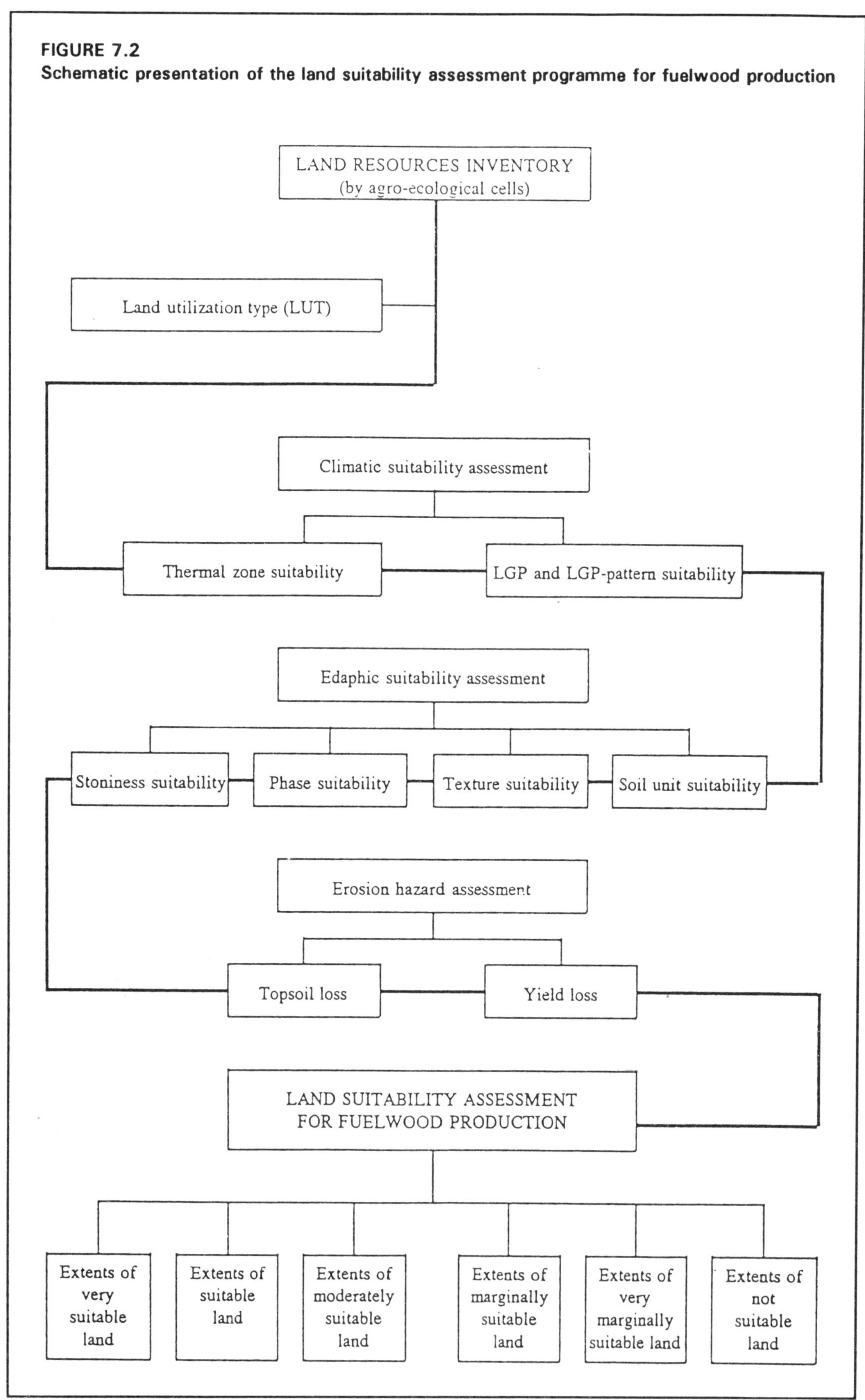

Step three relates the estimated topsoil losses to yield losses through a set of equations given in Table 4.4, taking into account soil susceptibility (Table 4.3), level of inputs and regeneration capacity of topsoil (Table 4.2).

7.5 Land Suitability Assessment

7.5.1 Fuelwood Productivity Potential

All three assessments: the climatic suitability, the edaphic suitability and the soil erosion hazard, are required to determine the ecological land suitability for fuelwood production of each climate-soil unit of the land resources inventory. In essence the land suitability assessment takes account of all the inventoried attributes of land and compares them with the requirements of the fuelwood species, to give an easy to understand picture of the suitability of land for the production of fuelwood.

The results of the land suitability assessment can be presented using five basic suitability classes, each linked to attainable yields (mean annual increment) for the three levels of inputs considered. For each level of inputs, the land suitability classes are: very suitable (VS) - 80% or more of the maximum attainable yield; suitable (S) - 60% to less than 80% of the maximum attainable yield; moderately suitable (MS) - 40% to less than 60% of the maximum attainable yields; marginally suitable (mS) - 20% to less than 40%; and not suitable (NS) - less than 20%.

Land suitability assessment for fuelwood production can be achieved by applying the programme illustrated in Figure 7.2. The assessment is carried out separately for each species and level of inputs, as explained in Section 5.1.5.

The five classes of land suitabilities are related to attainable yield as a percentage of the maximum attainable under the optimum climatic, edaphic and landform conditions, and so provide an assessment of fuelwood productivity potential of each land unit for calculation of the rainfed production potential of any given area in Kenya.

The generalized results of land suitability assessment for *Eucalyptus camaldulensis* at intermediate level of inputs are presented in Figure 7.3. Lands suitability results for all fuelwood species are presented in Technical Annex 8. It should be noted that the generalized results presented, include a subdivision of the not suitable class (zero to less than 20% of maximum attainable yield) into two classes (1) very marginally suitable (more than zero to less than 20% of maximum attainable yield) and (2) not suitable (zero yield).

7.5.2 Interphase with Crop and Livestock Productivity Models

Within the overall population supporting capacity model, the fuelwood productivity model is interphased with crop and livestock productivity models. The interphase in essence allows the possibility of considering:

(a) fuelwood production on land assessed as not suitable for crops;

(b) any portion of crop land for fuelwood production depending on how much land is required for other land uses to meet demand; and

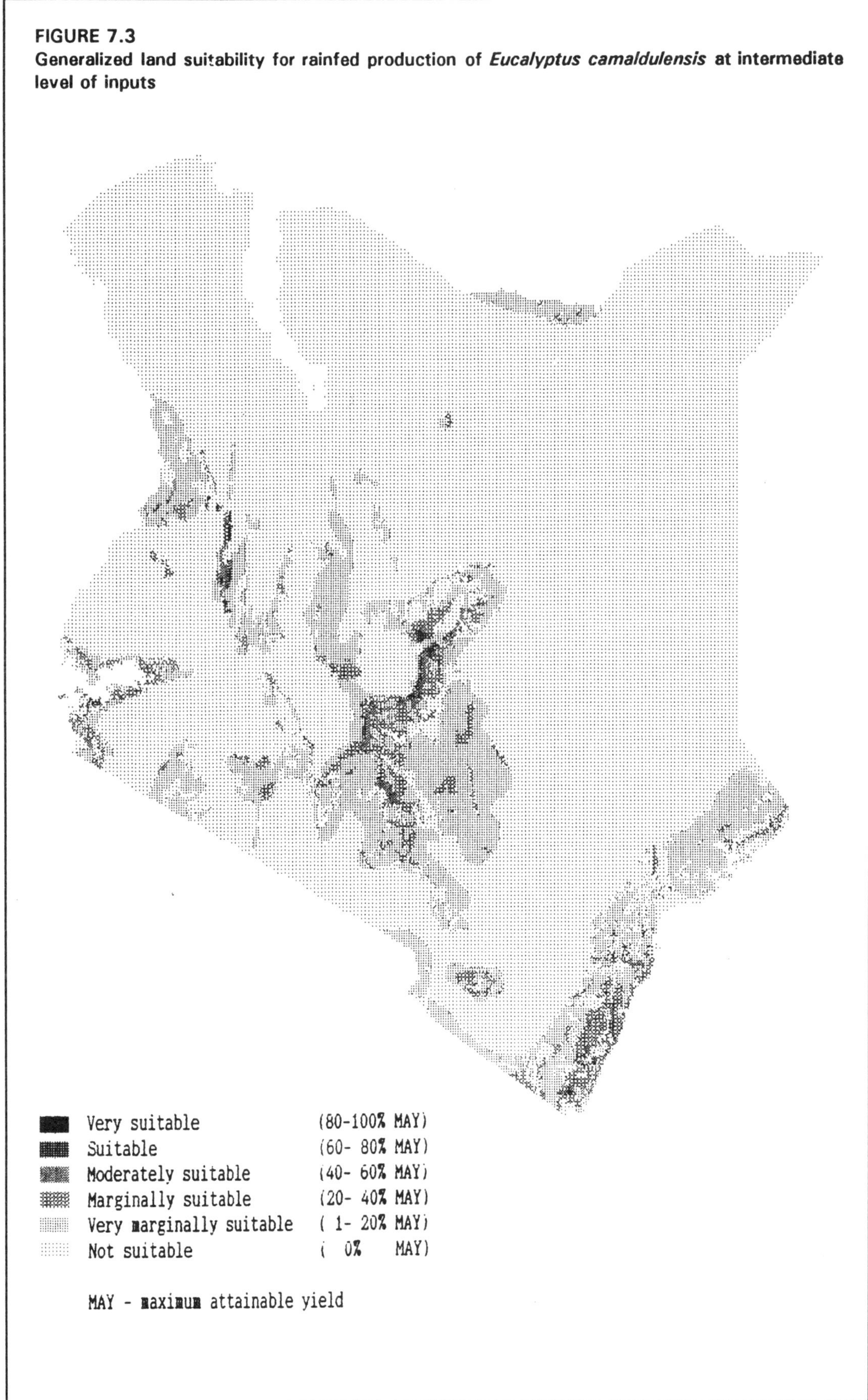

FIGURE 7.3
Generalized land suitability for rainfed production of *Eucalyptus camaldulensis* at intermediate level of inputs

(c) fodder from fuelwood trees for livestock production.

Any land which is allocated to fuelwood production with species that offers palatable foliage would have the potential of contributing a portion of this foliage to fodder for livestock production.

Fuelwood species which offer palatable fodder for livestock are listed in Table 7.1. The amount of fodder which can be utilized by stock without affecting fuelwood yields would depend on the species and ecological situation. However, at this stage of the model interphase developement, it is assumed that about 10 % of the foliage may be utilized by stock without affecting fuelwood yields. The nominal foliage utilization coefficient values may be modified as appropriate for each species and environment.

References

Andrews, D.J., & Kassam, A.H., 1976. Importance of multiple cropping in increasing world food supplies. pp. 1-10. *In: Multiple Cropping*. ASA Special Publication, No. 27.

Barr, D.A., 1957. The effect of sheet erosion on wheat yield. *J. Soil Conservation Service of New South Wales*, **13**(1): 27-32.

Blair Rains, A. & Kassam, A.H., 1980. Land resources and animal production [in Africa]. Working Paper No. 8, *In*: FAO 1980, *q.v.*

Boonman, J.G., 1979. Development of cultivated pastures in the highlands of East Africa. Nairobi: Ministry of Agriculture.

Boudet, G., & Riviere, R., 1968. Emploi practique des analyses fourrageres pour la l'appreciation des pasturages tropicaux. *Revue de l'Elevage et de Medecine Veterinaire des Pays Tropicaux*, **211**: 227-266.

Brammer, H., Antoine, J., Kassam, A.H, & van Velthuizen, H.T., 1988. Land resources appraisal of Bangladesh for agricultural development. Technical Reports Nos. 1-7, FAO/UNDP project BGD/81/035, 'Agricultural development advice'. Dhaka: Bangladesh.

Bruinsma, J., Hrabovszky, J., Alexandratos, N., & Petri, P., 1983. Crop production and input requirements in developing countries. *European Review of Agricultural Ecology*, **10**: 197-222.

Cannell, M.G.R., & Last, F.T., 1976. *Tree Physiology and Yield Improvement*. London: Academic Press.

Colvin, T.S., & Laflen, J.M., 1981. Effect of corn and soybean row spacing on canopy, erosion and runoff. *Trans. ASEA*, **24**(5): 1227-1229.

CRIES [Comprehensive Resource Inventory and Evaluation System], 1983. Kenya Natural Resource Assessment. The Comprehensive Resource Inventory and Evaluation System (CRIES) Project. Michigan State University.

Critchley, W., 1984. Runoff harvesting for crop range and tree production in the BPSAAP [Baringo Pilot Semi-arid and Arid Lands Project] Area. Draft report BPSAAP. Marigat, Kenya: Ministry of Agriculture and Livestock Development.

Davidson, J., 1985. Species and site: what to plant and where to plant. UNDP/FAO project BGD/79/017, 'Assistance to the forestry sector of Bangladesh'. Field Doc. No. 5. Dhaka.

Doorenbos, J., & Kassam, A.H., 1979. Yield Response to Water. *[FAO] Irrigation and Drainage Paper*, **33**. Rome.

Dunne, T., Dietrich, W.E., & Brunengo, M.J., 1978. Recent and past rates of erosion in semi-arid Kenya. *Zeitschrift fur Geomorphologie*, **29**(Suppl.): 99-100.

Edwards, D.C., & Bogdan, A.V., 1968. *Importance of grassland plants of Kenya*. Nairobi: Pitman and Sons.

FAO, 1974. *FAO-Unesco Soil Map of the World.* Vol. 1. *Legend.* Paris: Unesco.

FAO, 1976. A Framework for Land Evaluation. *FAO Soils Bulletin*, **32**.

FAO, 1977. Agroclimatological data bank. Rome, FAO/AGPC.

FAO, 1978-81. Report on the Agro-ecological Zones Project. Vol.1, Methodology and results for Africa; Vol.2, Results for Southwest Asia; Vol.3, Methodology and results for South and Central America; Vol.4, Results for Southeast Asia. *[FAO] World Soil Resources Report*, **48**/1,-4.

FAO, 1980. Report on the Second FAO/UNFPA Expert Consultation on Land Resources for Populations of the Future. Rome, FAO.

FAO, 1981. Eucalypts for planting. *FAO Forestry Series*, **11**.

FAO, 1982. Potential population supporting capacities of lands in the developing world. Technical Report of project INT/75/P13, 'Land resources for populations of the future', undertaken by FAO/IIASA for UNFPA.

FAO, 1984a. Land, Food and People. *FAO Econ. Soc. Dev. Series*, **30**.

FAO, 1984b. Population supporting capacity assessment of Kenya. Mission Report. Rome, AGL-FAO.

FAO, 1988a. FAO guidelines: land evaluation for extensive grazing. *FAO Soils Bulletin, 58*, draft edition.

FAO, 1988b. Pig Production in the Tropics. Proceedings of a seminar. Suchow, China. 21-25 September 1987. Edited N.R. Standal. Rome: FAO.

FAO, 1991. Guidelines: land evaluation for extensive grazing. *FAO Soils Bulletin*, **58**.

FAO/IIASA, 1991. Agro-ecological land resources assessment for agricultural development planning: A case study of Kenya: Resources database and land productivity. Main Report and 8 Technical Annexes. Rome, AGL-FAO. 9 vols. 1150p.

FAO/UNFPA, 1980. Report on the Second FAO/UNFPA Expert Consultation on Land Resources for Populations of the Future. Rome, Italy, 4-6 December 1979. Rome: FAO. 379 p.

Goehl, B., 1981. Tropical feeds. *FAO Animal Production and Health Series*, **12**.

Hammer, W.I., 1981. Soil Conservation Consultancy Report. Technical Note No.7, Centre for Soil Research, Bogor, Indonesia.

Higgins, G.M., & Kassam, A.H., 1981. Relating potential productivity to soil base. *[FAO] Land and Water Development Division Technical Newsletter*, **9** (July 1981).

Houerou, H.N. Le, & Hoste, C.H., 1977. Rangeland production and annual rainfall relations in the Mediterranean Basin and the African Sahelo-Sudanian Zone. *Journal of Range Management*, **30**(3): 181-189.

Hudson, N.W., 1981. *Soil Conservation*. London: Batsford.

Jaetzold, R. & Kutsch, H., 1980. Climatic data bank of Kenya. Department of Cultural and Regional Geography, University of Trier, Germany.

Jaetzold, R., & Schmidt, H., 1982. *Farm Management Handbook of Kenya*. Vol. II, Parts A, B & C. Nairobi: Ministry of Agriculture in cooperation with German Agency of Technical Cooperation.

Kassam, A.H., 1977. Net biomass production and yields of crops. Consultant's Report. Agroecological Zones project. FAO-AGL.

Kassam, A.H., 1980. Multiple cropping and rainfed crop productivity in Africa. pp. 123-195, *In*: FAO 1980, *q.v.*

Kassam, A.H., Kowal, J.M., & Sarraf, S., 1977. Climatic adaptability of crops. Consultants' Report. Agroecological Zones project. FAO-AGL, Rome.

Kassam, A.H., van Velthuizen, H.T., Higgins, G.M., Christoforides, A., Voortman, R.L., & Spiers, B., 1982. Assessment of land resources for rainfed crop production in Mozambique. Field Documents Nos. 32-37, FAO/UNDP project MOZ/75/011, 'Land and Water Use Planning'. Maputo, Mozambique.

King, J.M., Sayers, A.F., Peacock, C.P., & Kontrohr, E., 1982. Maasai herd and flock structure in relation to household livestock wealth and group ranch developement. *ILCA Working Document*, **27**.

Kowal, J.M., & Kassam, A.H., eds., 1978. *Agricultural Ecology of Savanna*. Oxford: Clarendon Press.

KSS [Kenya Soil Survey], 1975. Soils of the Kindaruma Area (Weg, R.F. van de, & Mbuvi, J.P., eds.). Nairobi: Kenya Soil Survey.

KSS, 1976. Soils of the Kapenguria Area (Weg, R.F. van de, ed.). Nairobi: Kenya Soil Survey.

KSS, 1982a. Exploratory Soil Map and Agroclimatic Zone Map of Kenya. Eds. Sombroek, W.G., Braun, H.M.H., & van der Pouw, B.J.A. Nairobi: Kenya Soil Survey.

KSS, 1982b. Soils of the Kisii Area (Wielemaker, W.G., & Boxem, H., eds.). Nairobi: Kenya Soil Survey.

Lal, R., 1976a. Soil erosion on alfisols in Western Nigeria, I: Effects of slope, crop rotation and residue management. *Geoderma*, **16**(5): 363-375.

Lal, R., 1976b. Soil erosion on alfisols in Western Nigeria, III: Effects of rainfall characteristics. *Geoderma*, **16**(5): 389-401.

Lal, R., 1976c. Soil erosion on alfisols in Western Nigeria, V: The changes in physical properties and the response of crops. *Geoderma*, **16**(5): 419-431.

Landsberg, J.J., 1986. *Physiological Ecology of Forest Production*. London: Academic Press.

Leeuw, P.N. de, & Peacock, C.P., 1982. The productivity of small ruminants in the Maasai Pastoral System in Kajiado District of Kenya. *ILCA Working Document*, **25**.

Meadows, S.J., & White, J.M., 1981. Cattle structure in Kenya's pastoral rangelands. *ODI Pastoral Network Paper*, **11e**.

Mitchell, A.J., 1986. Soil erosion and soil conservation. Consultancy Report. FAO-AGLS, Rome.

Nilsson, N.E., 1983. An alley model for forest resources planning. Statistics in Theory and Practice. Umea: Dept. Biometry and Forest Management, Swedish Univ. Agr. Sciences.

Nye, P.H., & Greenland, D.J., 1960. *The Soil Under Shifting Cultivation*. Bucks: Commonwealth Agricultural Bureaux.

Peacock, C.P., 1983. A rapid appraisal of goat and sheep flock demography in East and West Africa. Method, results and application to livestock research and development. *ILCA Working Document*, **28**.

Peacock, C.P., 1984. Sheep and goat production on Maasai group ranches. PhD Thesis. University of Reading, UK.

Pratt, D.J., & Gwynne, M.D., 1977. *Rangeland Management and Ecology in East Africa*. London: Hodder and Stoughton.

Rattray, J.M., 1960. *The grass cover of Africa*. Rome: FAO.

Sanchez, P.A., Couto, W., & Buol, S.W., 1982. The fertility capability soil classification system: interpretation, application and modification. *Geoderma*, **27**: 283-309.

Semenye, P.P., 1982. A preliminary report of cattle productivity in Olkarkar, Merueshi and Mbirikani group ranches. ILCA.

Siderius, W., & van der Pouw, B.J.A., 1980. The application of the FAO-Unesco terminology of the Soil Map of the World Legend for soil classification in Kenya. *Kenya Soil Survey Miscellaneous Soil Paper*, No. **M15**.

Skerman, P.J., 1982. Tropical Forage Legumes [1st edition]. *FAO Plant Production and Protection Series*, No. 2.

Stallings, J.H., 1957. *Soil Conservation*. Englewood Cliffs, NJ: Prentice-Hall.

Stotz, D., 1983. Production Techniques and Economics of Smallholder Livestock Production Systems in Kenya. Farm Management Handbook of Kenya, Vol. IV. Nairobi: Ministry of Agriculture.

Teel, W., 1980. *A Pocket Directory of Trees and Shrubs in Kenya*. Nairobi: KENGO.

Unesco, 1982. Resource management for the Rendille Area of Northern Kenya. Integrated Project on Arid Lands (IPAL), Marsabit.

Van der Pouw, B.J.A., 1983. Detailed composition of the soil mapping units of the Exploratory Soil Map of Kenya at scale 1:1 million. Wageningen: Stiboka.

Webb, D.B., Wood, P.J., & Smith, J., 1980. A guide to species selection for tropical and subtropical plantations. [Commonwealth Forestry Institute, Oxford] *Tropical Forestry Paper*, **15**.

Wiggins, S.L., & Palma, O.G., 1989. Cost benefit analysis of soil conservation. Land Resources Development Centre [Tolworth, UK] Project Report, 104.

Wischmeier, W.H., & Smith, D.D., 1978. Predicting rainfall erosion losses. United States Department of Agriculture Handbook, No. 536.

Young, A., & Wright, A.C.S., 1980. Rest period requirements of tropical and subtropical soils under annual crops. pp. *In:* FAO 1980, *q.v.*

Appendix

TABLE A5.1
Crop yields (t/ha dry weight) at the high input level by LGPs (days)

Crop	Growth cycle (days)	Max. yield	LGP 0	1-29	30-59	60-89	90-119	120-149	150-179	180-209	210-239	240-269	270-299	300-329	330-364	365⁻	365⁺
Barley	90-120	4.55	0.00	0.00	0.00	0.60	3.10	4.50	4.55	4.40	2.50	1.85	0.55	0.55	0.55	0.55	0.55
Barley	120-150	4.55	0.00	0.00	0.00	0.00	0.00	3.10	4.50	4.55	2.50	1.85	0.55	0.55	0.55	0.55	0.55
Barley	150-180	4.55	0.00	0.00	0.00	0.00	0.00	0.00	3.10	4.50	4.55	1.85	0.55	0.55	0.55	0.55	0.55
Maize (lowland)	70-90	4.14	0.00	0.00	0.00	0.67	1.63	3.11	4.14	4.04	2.81	1.92	1.42	1.32	0.99	0.99	0.40
Maize (lowland)	90-110	5.62	0.00	0.00	0.00	0.00	2.18	4.25	5.62	5.52	3.96	2.66	1.96	1.86	1.40	1.40	0.60
Maize (lowland)	110-130	7.10	0.00	0.00	0.00	0.00	0.00	5.40	7.10	7.00	5.10	3.40	2.50	2.40	1.80	1.80	0.80
Maize (highland)	120-140	8.50	0.00	0.00	0.00	0.23	1.78	5.89	8.50	7.80	5.85	3.90	2.93	2.93	2.19	0.98	0.98
Maize (highland)	140-180	9.40	0.00	0.00	0.00	0.00	0.00	0.00	7.09	8.37	9.40	8.26	6.22	3.40	2.55	1.14	1.14
Maize (highland)	180-200	10.30	0.00	0.00	0.00	0.00	0.00	0.00	3.69	6.59	10.30	10.30	7.73	3.86	2.90	1.29	1.29
Maize (highland)	200-220	10.90	0.00	0.00	0.00	0.00	0.00	0.00	3.69	7.13	10.90	10.90	8.18	4.09	3.07	1.36	1.36
Maize (highland)	220-280	10.90	0.00	0.00	0.00	0.00	0.00	0.00	2.09	3.96	6.51	7.02	7.19	5.65	3.34	1.48	1.48
Maize (highland)	280-300	10.90	0.00	0.00	0.00	0.00	0.00	0.00	0.49	0.78	2.12	3.13	6.19	7.20	3.60	1.60	1.60
Oats	90-120	4.10	0.00	0.00	0.00	0.50	2.80	4.10	4.00	3.90	2.20	1.60	0.45	0.45	0.45	0.45	0.45
Oats	120-150	4.10	0.00	0.00	0.00	0.00	0.00	2.80	4.10	4.00	2.20	1.60	0.45	0.45	0.45	0.45	0.45
Oats	150-180	4.10	0.00	0.00	0.00	0.00	0.00	0.00	2.80	4.10	4.00	1.60	0.45	0.45	0.45	0.45	0.45
Pearl millet	60-80	2.80	0.00	0.00	0.00	1.10	2.20	2.20	2.80	2.00	1.10	0.30	0.30	0.30	0.30	0.30	0.30
Pearl millet	80-100	3.90	0.00	0.00	0.00	0.00	3.10	3.10	3.90	2.80	1.60	0.50	0.50	0.50	0.50	0.50	0.50
Rice (dryland)	90-110	3.30	0.00	0.00	0.00	0.20	0.71	0.90	1.30	1.90	2.50	3.30	3.30	3.30	2.40	1.80	1.20
Rice (dryland)	110-130	4.20	0.00	0.00	0.00	0.00	0.00	1.10	1.60	2.40	3.20	4.20	4.20	4.10	3.00	2.30	1.50
Rice (wetland)	80-100	2.49	0.00	0.00	0.00	0.00	0.00	0.00	1.31	1.93	2.49	2.49	2.44	2.42	2.36	2.36	1.58
Rice (wetland)	100-120	3.28	0.00	0.00	0.00	0.00	0.00	0.00	1.71	2.52	3.28	3.28	3.23	3.20	3.15	3.15	2.10
Rice (wetland)	120-140	4.04	0.00	0.00	0.00	0.00	0.00	0.00	2.09	3.09	4.04	4.04	3.96	3.86	3.86	3.86	2.57
Sorghum (lowland)	70-90	2.97	0.00	0.00	0.00	0.51	1.52	2.33	2.97	2.87	1.50	0.67	0.50	0.30	0.30	0.30	0.30
Sorghum (lowland)	90-110	4.04	0.00	0.00	0.00	0.00	2.08	3.70	4.04	3.94	2.10	0.94	0.70	0.45	0.45	0.45	0.45
Sorghum (lowland)	110-130	5.10	0.00	0.00	0.00	0.00	0.00	3.80	5.10	5.00	2.70	1.20	0.90	0.60	0.60	0.60	0.60
Sorghum (highland)	120-140	6.00	0.00	0.00	0.00	0.23	1.24	4.20	6.00	5.60	3.15	1.40	1.05	0.70	0.70	0.70	0.70
Sorghum (highland)	140-180	6.65	0.00	0.00	0.00	0.00	0.00	0.00	5.31	6.65	6.46	4.25	2.35	1.28	0.81	0.81	0.81
Sorghum (highland)	180-200	7.30	0.00	0.00	0.00	0.00	0.00	0.00	2.59	4.68	7.30	5.48	2.74	1.37	0.91	0.91	0.91
Sorghum (highland)	200-220	7.80	0.00	0.00	0.00	0.00	0.00	0.00	2.59	5.14	7.80	5.85	2.93	1.46	0.98	0.98	0.98
Sorghum (highland)	220-280	7.80	0.00	0.00	0.00	0.00	0.00	0.00	1.47	2.94	4.61	4.39	3.65	2.46	1.26	1.26	1.26
Sorghum (highland)	280-300	7.80	0.00	0.00	0.00	0.00	0.00	0.00	0.35	0.74	1.42	2.93	4.36	3.45	1.53	1.53	1.53
Wheat	100-130	5.10	0.00	0.00	0.00	0.43	1.76	3.76	5.10	5.00	2.82	2.07	0.62	0.62	0.62	0.62	0.62
Wheat	130-160	5.00	0.00	0.00	0.00	0.00	0.00	3.12	5.00	5.00	2.82	2.07	0.62	0.62	0.62	0.62	0.62
Wheat	160-190	5.00	0.00	0.00	0.00	0.00	0.00	0.00	3.23	5.00	5.00	2.07	0.92	0.62	0.62	0.62	0.62

Crop	Growth cycle (days)	Max. yield	LGP 0	1-29	30-59	60-89	90-119	120-149	150-179	180-209	210-239	240-269	270-299	300-329	330-364	365^-	365^+
Cowpea	80-100	2.40	0.00	0.00	0.05	0.35	0.90	1.50	2.40	2.40	2.30	1.70	1.70	1.20	0.60	0.40	0.30
Cowpea	100-140	3.40	0.00	0.00	0.00	0.00	1.10	1.92	3.40	3.30	3.30	2.50	2.40	1.70	0.90	0.60	0.40
Green gram	60-80	1.90	0.00	0.00	0.08	0.27	1.00	1.40	1.90	1.90	1.80	1.40	1.40	0.60	0.50	0.40	0.30
Green gram	80-100	2.50	0.00	0.00	0.00	0.40	1.30	1.90	2.50	2.50	2.50	1.90	1.70	0.80	0.60	0.60	0.40
Groundnut	80-100	2.40	0.00	0.00	0.00	0.41	0.90	1.50	2.40	2.40	2.30	1.30	1.00	0.80	0.60	0.40	0.30
Groundnut	100-140	3.30	0.00	0.00	0.00	0.00	1.10	1.90	3.30	3.30	3.30	1.90	1.40	1.20	0.90	0.60	0.40
Phaseoulus bean	90-120	2.65	0.00	0.00	0.00	0.32	0.92	2.00	2.65	2.55	1.92	1.92	0.96	0.72	0.48	0.32	0.32
Phaseoulus bean	120-150	2.65	0.00	0.00	0.00	0.00	0.37	1.00	2.65	2.55	1.92	1.92	0.96	0.72	0.48	0.32	0.32
Phaseoulus bean	150-180	2.65	0.00	0.00	0.00	0.00	0.00	0.44	1.49	2.65	2.55	1.92	0.96	0.72	0.48	0.32	0.32
Pigeon pea	130-150	3.30	0.00	0.00	0.00	0.07	0.35	1.38	3.30	3.30	3.20	2.40	2.40	1.70	0.90	0.60	0.40
Pigeon pea	150-170	3.80	0.00	0.00	0.00	0.00	0.00	1.50	3.80	3.70	3.70	2.80	2.70	2.00	1.00	0.70	0.40
Pigeon pea	170-190	4.04	0.00	0.00	0.00	0.00	0.00	0.00	3.55	3.98	4.04	3.08	2.91	2.21	1.07	0.77	0.40
Soybean	80-100	2.40	0.00	0.00	0.00	0.35	0.90	1.90	2.40	2.40	1.70	1.10	0.90	0.60	0.40	0.30	0.30
Soybean	100-140	3.40	0.00	0.00	0.00	0.00	1.00	2.50	3.40	3.30	2.50	1.60	1.20	0.90	0.60	0.40	0.40
Cassava	150-330	13.15	0.00	0.00	0.00	0.00	0.00	0.00	3.50	8.75	10.85	11.90	12.40	13.15	7.40	7.40	5.00
Sweet potato	115-125	7.70	0.00	0.00	0.00	0.00	2.05	3.90	7.60	7.60	7.40	7.40	5.50	2.70	2.70	2.70	1.80
Sweet potato	125-145	8.90	0.00	0.00	0.00	0.00	0.00	4.50	8.75	8.75	8.55	8.55	6.30	3.10	3.05	3.05	2.05
Sweet potato	145-155	10.10	0.00	0.00	0.00	0.00	0.00	0.00	10.10	9.90	9.70	9.70	7.10	3.50	3.40	3.40	2.30
White potato	90-110	5.70	0.00	0.00	0.00	1.20	2.08	3.26	5.70	5.48	3.05	3.05	1.05	0.70	0.70	0.70	0.70
White potato	110-130	7.30	0.00	0.00	0.00	0.00	2.25	4.15	7.30	7.05	3.95	3.95	1.35	0.90	0.90	0.90	0.90
White potato	130-170	9.70	0.00	0.00	0.00	0.00	0.00	4.84	9.70	9.40	5.30	5.30	1.80	1.20	1.20	1.20	1.20
Banana	300-365	9.60[1]	0.00	0.00	0.00	0.00	0.00	0.00	0.00	0.00	0.00	0.00	2.30	4.30	6.70	9.60	9.60
Oil palm	270-365	5.30[2]	0.00	0.00	0.00	0.00	0.00	0.00	0.00	0.00	0.00	0.00	2.81	3.28	4.44	5.30	5.30
Sugarcane	210-365	11.60[1]	0.00	0.00	0.00	0.00	0.00	0.00	0.00	0.00	4.40	6.30	8.10	9.80	10.82	11.60	8.70

1 Adjusted as appropriate for turn-round time (15 days) between crops where LGP >330.
2 Adjusted for the non-productive establishment period (20% of total rotation age).

TABLE A5.2
Crop yields (t/ha dry weight) at the intermediate input level by LGPs (days)

Crop	Growth cycle (days)	Max. yield	LGP														
			0	1-29	30-59	60-89	90-119	120-149	150-179	180-209	210-239	240-269	270-299	300-329	330-364	365⁻	365⁺
Barley	90-120	2.85	0.00	0.00	0.00	0.38	1.93	2.83	2.85	2.75	1.55	1.15	0.38	0.38	0.35	0.35	0.35
Barley	120-150	2.85	0.00	0.00	0.00	0.00	0.00	1.93	2.83	2.85	1.68	1.25	0.38	0.38	0.35	0.35	0.35
Barley	150-180	2.85	0.00	0.00	0.00	0.00	0.00	0.00	1.93	2.83	2.85	1.35	0.38	0.38	0.35	0.35	0.35
Maize (lowland)	70-90	2.59	0.00	0.00	0.00	0.36	1.00	1.86	2.57	2.54	1.79	1.26	1.03	0.91	0.70	0.58	0.29
Maize (lowland)	90-110	3.51	0.00	0.00	0.00	0.00	1.30	2.53	3.51	3.45	2.50	1.73	1.37	1.28	0.99	0.82	0.42
Maize (lowland)	110-130	4.45	0.00	0.00	0.00	0.00	0.00	3.20	4.45	4.35	3.20	2.20	1.70	1.65	1.25	1.05	0.55
Maize (highland)	120-140	5.32	0.00	0.00	0.00	0.14	1.06	3.50	5.32	4.88	3.66	2.50	2.02	2.02	1.37	0.61	0.61
Maize (highland)	140-180	5.84	0.00	0.00	0.00	0.00	0.00	0.00	3.55	5.20	5.84	5.17	3.97	2.34	1.58	0.71	0.71
Maize (highland)	180-200	6.44	0.00	0.00	0.00	0.00	0.00	0.00	2.31	4.12	6.44	6.44	4.83	2.66	1.81	0.81	0.81
Maize (highland)	200-220	6.82	0.00	0.00	0.00	0.00	0.00	0.00	2.31	4.56	6.82	6.82	5.11	2.81	1.92	0.85	0.85
Maize (highland)	220-280	6.82	0.00	0.00	0.00	0.00	0.00	0.00	1.31	7.23	4.07	4.39	4.49	3.66	2.09	0.93	0.93
Maize (highland)	280-300	6.82	0.00	0.00	0.00	0.00	0.00	0.00	0.31	0.49	3.71	3.96	3.87	4.50	2.25	1.00	1.00
Oats	90-120	2.55	0.00	0.00	0.00	0.33	1.75	2.55	2.48	2.43	1.38	0.98	0.30	0.30	0.28	0.28	0.28
Oats	120-150	2.55	0.00	0.00	0.00	0.00	0.00	1.75	2.55	2.43	1.58	1.08	0.30	0.30	0.28	0.28	0.28
Oats	150-180	2.55	0.00	0.00	0.00	0.00	0.00	0.00	1.75	2.55	2.48	1.28	0.30	0.30	0.28	0.28	0.28
Pearl millet	60-80	1.75	0.00	0.00	0.00	0.70	1.35	1.35	1.75	1.25	0.75	0.25	0.20	0.20	0.20	0.20	0.20
Pearl millet	80-100	2.45	0.00	0.00	0.00	0.00	1.95	1.95	2.45	1.75	1.05	0.35	0.35	0.30	0.30	0.30	0.30
Rice (dryland)	90-110	2.07	0.00	0.00	0.00	0.13	0.45	0.57	0.82	1.19	1.57	2.07	2.07	2.07	1.50	1.13	0.75
Rice (dryland)	110-130	2.63	0.00	0.00	0.00	0.00	0.00	0.69	1.00	1.50	2.00	2.63	2.63	2.57	1.88	1.44	0.94
Rice (wetland)	80-100	1.56	0.00	0.00	0.00	0.00	0.00	0.00	0.79	1.21	1.56	1.56	1.53	1.52	1.48	1.48	0.99
Rice (wetland)	100-120	2.05	0.00	0.00	0.00	0.00	0.00	0.00	1.07	1.58	2.05	2.03	2.02	2.00	1.97	1.97	1.32
Rice (wetland)	120-140	2.53	0.00	0.00	0.00	0.00	0.00	0.00	1.31	1.93	2.53	2.53	2.48	2.47	2.42	2.42	1.61
Sorghum (lowland)	70-90	1.86	0.00	0.00	0.00	0.31	0.91	1.39	1.86	1.80	0.94	0.42	0.30	0.18	0.18	0.18	0.18
Sorghum (lowland)	90-110	2.53	0.00	0.00	0.00	0.00	1.24	2.14	2.51	1.97	1.32	0.59	0.42	0.27	0.27	0.27	0.27
Sorghum (lowland)	110-130	3.18	0.00	0.00	0.00	0.00	0.00	2.26	3.19	3.13	1.69	0.75	0.54	0.36	0.36	0.36	0.36
Sorghum (highland)	120-140	3.75	0.00	0.00	0.00	0.14	0.74	2.50	3.75	3.50	1.97	0.90	0.66	0.48	0.44	0.44	0.44
Sorghum (highland)	140-180	4.15	0.00	0.00	0.00	0.00	0.00	0.00	3.35	4.15	4.07	2.69	1.51	0.79	0.51	0.51	0.51
Sorghum (highland)	180-200	4.57	0.00	0.00	0.00	0.00	0.00	0.00	1.62	2.93	4.57	3.43	1.76	0.86	0.57	0.57	0.57
Sorghum (highland)	200-220	4.88	0.00	0.00	0.00	0.00	0.00	0.00	1.62	3.22	4.88	3.66	1.88	0.92	0.61	0.62	0.62
Sorghum (highland)	220-280	4.88	0.00	0.00	0.00	0.00	0.00	0.00	0.91	1.84	2.89	2.75	2.24	1.65	0.79	0.79	0.79
Sorghum (highland)	280-300	4.88	0.00	0.00	0.00	0.00	0.00	0.00	0.20	0.46	0.89	1.83	2.59	2.37	0.96	0.96	0.96
Wheat	100-130	3.19	0.00	0.00	0.00	0.27	1.12	2.35	3.19	3.13	1.77	1.30	0.43	0.43	0.39	0.39	0.39
Wheat	130-160	3.13	0.00	0.00	0.00	0.00	0.00	1.95	3.13	3.13	1.88	1.38	0.43	0.43	0.39	0.39	0.39
Wheat	160-190	3.13	0.00	0.00	0.00	0.00	0.00	0.00	2.02	3.13	3.13	1.38	0.64	0.43	0.39	0.39	0.39

Crop	Growth cycle (days)	Max. yield	LGP 0	1-29	30-59	60-89	90-119	120-149	150-179	180-209	210-239	240-269	270-299	300-329	330-364	365^-	365^+
Cowpea	80-100	1.45	0.00	0.00	0.03	0.20	0.50	0.85	1.40	1.45	1.45	1.00	1.00	0.75	0.40	0.30	0.25
Cowpea	100-140	1.95	0.00	0.00	0.00	0.00	0.60	1.11	1.95	1.95	1.95	1.50	1.45	1.10	0.65	0.40	0.30
Green gram	60-80	1.20	0.00	0.00	0.05	0.17	0.60	0.90	1.20	1.20	1.15	0.80	0.80	0.40	0.30	0.20	0.15
Green gram	80-100	1.55	0.00	0.00	0.00	0.25	0.80	1.20	1.55	1.55	1.55	1.10	1.00	0.50	0.35	0.25	0.20
Groundnut	80-100	1.50	0.00	0.00	0.00	0.28	0.55	0.95	1.50	1.50	1.35	0.80	0.60	0.50	0.40	0.25	0.20
Groundnut	100-140	2.05	0.00	0.00	0.00	0.00	0.70	1.20	2.05	2.05	1.95	1.20	0.90	0.70	0.55	0.40	0.25
Phaseoulus bean	90-120	1.66	0.00	0.00	0.00	0.20	0.58	1.25	1.66	1.52	1.14	1.05	0.57	0.45	0.28	0.20	0.20
Phaseoulus bean	120-150	1.66	0.00	0.00	0.00	0.00	0.23	0.63	1.66	1.52	1.14	1.08	0.57	0.45	0.28	0.20	0.20
Phaseoulus bean	150-180	1.66	0.00	0.00	0.00	0.00	0.00	0.28	0.93	1.57	1.46	1.08	0.57	0.45	0.28	0.20	0.20
Pigeon pea	130-150	2.05	0.00	0.00	0.00	0.05	0.22	0.87	2.05	2.05	2.00	1.50	1.40	1.05	0.55	0.35	0.25
Pigeon pea	150-170	2.40	0.00	0.00	0.00	0.00	0.00	0.95	2.40	2.30	2.30	1.75	1.60	1.25	0.60	0.45	0.25
Pigeon pea	170-190	2.53	0.00	0.00	0.00	0.00	0.00	0.00	2.22	2.49	2.53	1.93	1.82	1.33	0.67	0.49	0.25
Soybean	80-100	1.50	0.00	0.00	0.00	0.21	0.55	1.20	1.50	1.40	1.00	0.70	0.55	0.35	0.25	0.20	0.20
Soybean	100-140	2.10	0.00	0.00	0.00	0.00	0.63	1.55	2.10	1.95	1.50	1.05	0.75	0.55	0.35	0.25	0.25
Cassava	150-330	8.24	0.00	0.00	0.00	0.00	0.00	0.00	2.35	5.20	6.76	7.42	7.74	8.24	4.97	4.97	3.44
Sweet potato	115-125	4.80	0.00	0.00	0.00	0.00	1.28	2.45	4.80	4.75	4.60	4.40	3.45	2.05	1.80	1.80	1.25
Sweet potato	125-145	5.55	0.00	0.00	0.00	0.00	0.00	2.83	5.55	5.48	5.33	5.08	3.95	2.35	2.05	2.05	1.45
Sweet potato	145-155	6.30	0.00	0.00	0.00	0.00	0.00	0.00	6.30	6.20	6.05	5.75	4.45	2.65	2.30	2.30	1.60
White potato	90-110	3.55	0.00	0.00	0.00	0.73	1.29	2.04	3.55	3.44	2.05	1.93	0.71	0.42	0.42	0.42	0.42
White potato	110-130	4.55	0.00	0.00	0.00	0.00	1.40	2.60	4.55	4.43	2.65	2.48	0.93	0.55	0.55	0.55	0.55
White potato	130-170	6.05	0.00	0.00	0.00	0.00	0.00	3.12	6.05	5.90	3.55	3.30	1.25	0.75	0.75	0.75	0.75
Banana	300-365	6.00[1]	0.00	0.00	0.00	0.00	0.00	0.00	0.00	0.00	0.00	0.00	1.45	2.70	4.00	6.00	6.00
Oil palm	270-365	3.32[2]	0.00	0.00	0.00	0.00	0.00	0.00	0.00	0.00	0.00	0.00	1.69	2.06	2.78	3.32	3.32
Sugarcane	210-365	7.25[1]	0.00	0.00	0.00	0.00	0.00	0.00	0.00	0.00	2.75	3.95	5.05	6.10	6.77	7.26	5.44

1 Adjusted as appropriate for turn-round time (15 days) between crops where LGP >330.
2 Adjusted for the non-productive establishment period (20% of total rotation age).

TABLE A5.3
Crop yields (t/ha dry weight) at the low input level by LGPs (days)

Crop	Growth cycle (days)	Max. yield	LGP 0	1-29	30-59	60-89	90-119	120-149	150-179	180-209	210-239	240-269	270-299	300-329	330-364	365⁻	365⁺
Barley	90-120	1.15	0.00	0.00	0.00	0.15	0.75	1.15	1.15	1.10	0.06	0.45	0.20	0.20	0.15	0.15	0.15
Barley	120-150	1.15	0.00	0.00	0.00	0.00	0.00	0.75	1.15	1.15	0.85	0.65	0.20	0.20	0.15	0.15	0.15
Barley	150-180	1.15	0.00	0.00	0.00	0.00	0.00	0.00	1.15	1.15	1.15	0.85	0.20	0.20	0.20	0.15	0.35
Maize (lowland)	70-90	1.03	0.00	0.00	0.00	0.12	0.37	0.60	0.99	1.03	0.76	0.60	0.63	0.50	0.40	0.17	0.17
Maize (lowland)	90-110	1.40	0.00	0.00	0.00	0.00	0.41	0.80	1.40	1.37	1.03	0.80	0.77	0.70	0.55	0.24	0.24
Maize (lowland)	110-130	1.80	0.00	0.00	0.00	0.00	0.00	1.00	1.80	1.70	1.30	1.00	0.90	0.90	0.70	0.30	0.30
Maize (highland)	120-140	2.13	0.00	0.00	0.00	0.05	0.34	1.10	2.13	1.95	1.46	1.10	1.10	1.10	0.55	0.24	0.24
Maize (highland)	140-180	2.27	0.00	0.00	0.00	0.00	0.00	0.00	1.72	2.02	2.27	2.07	1.71	1.28	0.61	0.28	0.28
Maize (highland)	180-200	2.58	0.00	0.00	0.00	0.00	0.00	0.00	0.92	1.65	2.58	2.58	1.93	1.45	0.72	0.32	0.32
Maize (highland)	200-220	2.73	0.00	0.00	0.00	0.00	0.00	0.00	0.92	1.99	2.73	2.73	2.04	1.53	0.77	0.34	0.34
Maize (highland)	220-280	2.73	0.00	0.00	0.00	0.00	0.00	0.00	0.53	1.05	1.62	1.76	1.79	1.67	0.84	0.37	0.37
Maize (highland)	280-300	2.73	0.00	0.00	0.00	0.00	0.00	0.00	0.13	0.20	0.53	0.79	1.54	1.80	0.90	0.40	0.40
Oats	90-120	1.00	0.00	0.00	0.00	0.15	0.70	1.00	0.95	0.95	0.55	0.35	0.15	0.15	0.10	0.10	0.10
Oats	120-150	1.00	0.00	0.00	0.00	0.00	0.00	0.70	1.00	0.95	0.95	0.55	0.15	0.15	0.10	0.10	0.10
Oats	150-180	1.00	0.00	0.00	0.00	0.00	0.00	0.00	0.70	1.00	0.95	0.95	0.15	0.15	0.10	0.10	0.10
Pearl millet	60-80	0.70	0.00	0.00	0.00	0.30	0.50	0.50	0.70	0.50	0.40	0.20	0.10	0.10	0.10	0.10	0.10
Pearl millet	80-100	1.00	0.00	0.00	0.00	0.00	0.80	0.80	1.00	0.70	0.50	0.20	0.20	0.20	0.20	0.20	0.20
Rice (dryland)	90-110	0.83	0.00	0.00	0.00	0.05	0.18	0.23	0.33	0.48	0.63	0.83	0.83	0.83	0.60	0.45	0.30
Rice (dryland)	110-130	1.05	0.00	0.00	0.00	0.00	0.00	0.28	0.40	0.60	0.80	1.05	1.05	1.03	0.75	0.58	0.38
Rice (wetland)	80-100	0.62	0.00	0.00	0.00	0.00	0.00	0.00	0.26	0.48	0.62	0.62	0.61	0.61	0.59	0.59	0.40
Rice (wetland)	100-120	0.82	0.00	0.00	0.00	0.00	0.00	0.00	0.43	0.63	0.82	0.82	0.81	0.80	0.79	0.79	0.53
Rice (wetland)	120-140	1.01	0.00	0.00	0.00	0.00	0.00	0.00	0.52	0.77	1.01	1.01	0.99	0.99	0.97	0.97	0.64
Sorghum (lowland)	70-90	0.74	0.00	0.00	0.00	0.10	0.29	0.44	0.74	0.72	0.38	0.17	0.09	0.06	0.06	0.06	0.06
Sorghum (lowland)	90-110	1.01	0.00	0.00	0.00	0.00	0.39	0.58	1.01	0.99	0.53	0.24	0.13	0.08	0.08	0.08	0.08
Sorghum (lowland)	110-130	1.28	0.00	0.00	0.00	0.00	0.00	0.71	1.28	1.25	0.68	0.30	0.17	0.11	0.11	0.11	0.11
Sorghum (highland)	120-140	1.50	0.00	0.00	0.00	0.05	0.23	0.79	1.50	1.40	0.79	0.39	0.26	0.26	0.18	0.18	0.18
Sorghum (highland)	140-180	1.67	0.00	0.00	0.00	0.00	0.00	0.00	1.38	1.65	1.68	1.13	0.67	0.30	0.21	0.21	0.21
Sorghum (highland)	180-200	1.83	0.00	0.00	0.00	0.00	0.00	0.00	0.65	1.17	1.83	1.37	0.77	0.34	0.23	0.23	0.23
Sorghum (highland)	200-220	1.95	0.00	0.00	0.00	0.00	0.00	0.00	0.65	1.29	1.95	1.46	0.82	0.37	0.24	0.24	0.24
Sorghum (highland)	220-280	1.95	0.00	0.00	0.00	0.00	0.00	0.00	0.35	0.74	1.16	1.10	0.82	0.83	0.32	0.32	0.32
Sorghum (highland)	280-300	1.95	0.00	0.00	0.00	0.00	0.00	0.00	0.05	0.18	0.36	0.73	0.81	1.29	0.39	0.39	0.39
Wheat	100-130	1.28	0.00	0.00	0.00	0.11	0.47	0.94	1.28	1.25	0.71	0.52	0.23	0.23	0.15	0.15	0.15
Wheat	130-160	1.25	0.00	0.00	0.00	0.00	0.00	0.78	1.25	1.25	0.94	0.69	0.23	0.23	0.15	0.15	0.15
Wheat	160-190	1.25	0.00	0.00	0.00	0.00	0.00	0.00	0.81	1.25	1.25	0.69	0.35	0.23	0.15	0.15	0.15

Crop	Growth cycle (days)	Max. yield	LGP 0	1-29	30-59	60-89	90-119	120-149	150-179	180-209	210-239	240-269	270-299	300-329	330-364	365⁻	365⁺
Cowpea	80-100	0.50	0.00	0.00	0.00	0.05	0.10	0.20	0.40	0.50	0.50	0.30	0.30	0.30	0.20	0.20	0.20
Cowpea	100-140	0.60	0.00	0.00	0.00	0.00	0.10	0.30	0.50	0.60	0.60	0.50	0.50	0.50	0.40	0.20	0.20
Green gram	60-80	0.50	0.00	0.00	0.02	0.07	0.20	0.40	0.50	0.50	0.50	0.20	0.20	0.20	0.10	0.10	0.10
Green gram	80-100	0.60	0.00	0.00	0.00	0.10	0.30	0.50	0.60	0.60	0.60	0.30	0.30	0.20	0.10	0.10	0.10
Groundnut	80-100	0.60	0.00	0.00	0.00	0.14	0.20	0.40	0.60	0.60	0.40	0.30	0.20	0.20	0.20	0.10	0.10
Groundnut	100-140	0.80	0.00	0.00	0.00	0.00	0.30	0.50	0.80	0.80	0.60	0.50	0.40	0.20	0.20	0.20	0.10
Phaseoulus bean	90-120	0.67	0.00	0.00	0.00	0.07	0.23	0.56	0.67	0.48	0.36	0.18	0.18	0.18	0.08	0.08	0.08
Phaseoulus bean	120-150	0.67	0.00	0.00	0.00	0.00	0.09	0.25	0.67	0.48	0.36	0.24	0.18	0.18	0.08	0.08	0.08
Phaseoulus bean	150-180	0.67	0.00	0.00	0.00	0.00	0.00	0.11	0.37	0.48	0.36	0.24	0.18	0.18	0.08	0.08	0.08
Pigeon pea	130-150	0.80	0.00	0.00	0.00	0.03	0.08	0.36	0.80	0.80	0.80	0.60	0.40	0.40	0.20	0.10	0.10
Pigeon pea	150-170	1.00	0.00	0.00	0.00	0.00	0.00	0.40	1.00	0.90	0.90	0.70	0.50	0.50	0.20	0.20	0.10
Pigeon pea	170-190	1.01	0.00	0.00	0.00	0.00	0.00	0.00	0.89	1.00	1.01	0.77	0.73	0.55	0.27	0.20	0.10
Soybean	80-100	0.60	0.00	0.00	0.00	0.07	0.20	0.50	0.60	0.40	0.30	0.30	0.20	0.10	0.10	0.10	0.10
Soybean	100-140	0.80	0.00	0.00	0.00	0.00	0.25	0.60	0.80	0.60	0.50	0.50	0.30	0.20	0.10	0.10	0.10
Cassava	150-330	3.30	0.00	0.00	0.00	0.00	0.00	0.00	1.20	1.65	2.67	2.93	3.07	3.33	2.53	2.53	1.87
Sweet potato	115-125	1.90	0.00	0.00	0.00	0.00	0.50	1.00	1.90	1.90	1.80	1.40	1.40	1.40	0.90	0.90	0.70
Sweet potato	125-145	2.20	0.00	0.00	0.00	0.00	0.00	1.15	2.20	2.20	2.10	1.60	1.60	1.60	1.05	1.05	0.85
Sweet potato	145-155	2.50	0.00	0.00	0.00	0.00	0.00	0.00	2.50	2.50	2.40	1.80	1.80	1.80	1.20	1.20	0.90
White potato	90-110	1.40	0.00	0.00	0.00	0.26	0.50	0.82	1.40	1.40	1.05	0.80	0.37	0.13	0.13	0.13	0.13
White potato	110-130	1.80	0.00	0.00	0.00	0.00	0.55	1.05	1.80	1.80	1.35	1.00	0.50	0.20	0.20	0.20	0.20
White potato	130-170	2.40	0.00	0.00	0.00	0.00	0.00	1.40	2.40	2.40	1.80	1.30	0.70	0.30	0.30	0.30	0.30
Banana	300-365	2.40[1]	0.00	0.00	0.00	0.00	0.00	0.00	0.00	0.00	0.00	0.00	0.60	1.10	1.60	2.40	2.40
Oil palm	270-365	1.33[2]	0.00	0.00	0.00	0.00	0.00	0.00	0.00	0.00	0.00	0.00	0.56	0.82	1.11	1.33	1.33
Sugarcane	210-365	2.91[1]	0.00	0.00	0.00	0.00	0.00	0.00	0.00	0.00	1.10	1.60	2.00	2.40	2.71	2.91	2.18

1 Adjusted as appropriate for turn-round time (15 days) between crops where LGP >330.

2 Adjusted for the non-productive establishment period (20% of total rotation age).

TABLE A7.1
Total biomass yield potential (Bm) without constraints (mean annual increment in t/ha dry weight) at high level of inputs

Length of growing period (days)	Species without nitrogen fixing ability			Species with nitrogen fixing ability		
	Pm = 7.5	Pm = 15.0	Pm = 25.0	Pm = 7.5	Pm = 15.0	Pm = 25.0
1 - 29	0.0-0.3	0.0-0.4	0.0-0.6	0.0-0.2	0.0-0.4	0.0-0.6
30 - 59	0.3-0.2	0.4-3.3	0.6-4.7	0.2-1.7	0.4-2.9	0.6-4.0
60 - 89	2.0-4.2	3.3-7.1	4.7-10.0	1.7-3.4	2.9-5.8	4.0-8.1
90 - 119	4.2-6.2	7.1-10.6	10.0-14.9	3.4-4.8	5.8-8.2	8.1-11.5
120 - 149	6.2-9.0	10.6-15.4	14.9-21.6	4.8-6.6	8.2-11.3	11.5-16.0
150 - 179	9.0-11.0	15.4-18.7	21.6-26.3	6.6-7.8	11.3-13.3	16.0-18.7
180 - 209	11.0-13.6	18.7-23.3	26.3-32.8	7.8-9.4	13.3-16.0	18.7-22.5
210 - 239	13.6-15.0	23.3-25.5	32.8-36.0	9.4-10.0	16.0-17.0	28.7-22.5
240 - 269	15.0-16.2	25.5-27.7	36.0-39.0	10.0-10.5	17.0-17.9	22.5-24.0
270 - 299	16.2-17.4	27.7-29.6	39.0-41.8	10.5-11.0	17.9-18.7	25.3-26.4
300 - 329	17.4-18.5	29.6-31.5	41.8-44.4	11.0-11.4	18.7-19.4	26.4-27.4
330 - 364	18.5-19.6	31.5-33.5	44.4-47.2	11.4-11.8	19.4-20.2	27.4-28.5
365-	19.6	33.5	47.2	11.8	20.2	28.5
365+	19.6	33.5	47.2	11.8	20.2	28.5

Pm - maximum photosynthesis rate in kg CH_2O ha^{-1} hr^{-1}

TABLE A7.2
Total biomass yield potential (Bm) without constraints (mean annual increment in t/ha dry weight) at intermediate level of inputs

Length of growing period (days)	Species without nitrogen fixing ability			Species with nitrogen fixing ability		
	Pm = 7.5	Pm = 15.0	Pm = 25.0	Pm = 7.5	Pm = 15.0	Pm = 25.0
1 - 29	0.0-0.2	0.0-0.3	0.0-0.5	0.0-0.2	0.0-0.3	0.0-0.5
30 - 59	0.2-1.5	0.3-2.5	0.5-3.5	0.2-1.3	0.3-2.2	0.5-3.0
60 - 89	1.5-3.2	2.5-5.3	3.5-7.5	1.3-2.6	2.2-4.4	3.0-6.1
90 - 119	3.2-4.7	5.3-8.0	7.5-11.2	2.6-3.6	4.4-6.2	6.1-8.6
120 - 149	4.7-6.8	8.0-11.6	11.2-16.2	3.6-5.0	6.2-8.5	8.6-12.0
150 - 179	6.8-8.3	11.6-14.0	16.2-19.7	5.0-5.9	8.5-10.0	12.0-14.0
180 - 209	8.3-10.2	14.0-17.5	19.7-24.6	5.9-7.1	10.0-12.0	14.0-16.9
210 - 239	10.2-11.3	17.5-19.1	24.6-27.0	7.1-7.5	12.0-12.8	16.9-18.0
240 - 269	11.3-12.2	19.1-20.8	27.0-29.3	7.5-7.9	12.8-13.4	18.0-19.0
270 - 299	12.2-13.1	20.8-22.2	29.3-31.4	7.9-8.3	13.4-14.0	19.0-19.8
300 - 329	13.1-13.9	22.2-23.6	31.4-33.3	8.3-8.6	14.0-14.6	19.8-20.6
330 - 364	13.9-14.7	23.6-25.1	33.3-35.4	8.6-8.9	14.6-15.2	20.6-21.4
365-	14.7	25.1	35.4	8.9	15.2	21.4
365+	14.7	25.1	35.4	8.9	15.2	21.4

Pm - maximum photosynthesis rate in kg CH_2O ha^{-1} hr^{-1}

TABLE A7.3
Total biomass yield potential (Bm) without constraints (mean annual increment in t/ha dry weight) at low level of inputs

Length of growing period (days)	Species without nitrogen fixing ability			Species with nitrogen fixing ability		
	Pm=7.5	Pm=15.0	Pm=25.0	Pm=7.5	Pm=15.0	Pm=25.0
1 - 29	0.0-0.2	0.0-0.2	0.0-0.3	0.0-0.1	0.0-0.2	0.0-0.3
30 - 59	0.2-1.0	0.2-1.7	0.3-2.4	0.1-0.9	0.2-1.5	0.3-2.0
60 - 89	1.0-2.1	1.7-3.6	2.4-5.0	0.9-1.7	1.5-2.9	2.0-4.1
90 - 119	2.1-3.1	3.6-5.3	5.0-7.5	1.7-2.4	2.9-4.1	4.1-5.8
120 - 149	3.1-4.5	5.3-7.7	7.5-10.8	2.4-3.3	4.1-5.7	5.8-8.0
150 - 179	4.5-5.5	7.7-9.4	10.8-13.2	3.3-3.9	5.7-6.7	8.0-9.4
180 - 209	5.5-6.8	9.4-11.7	13.2-16.4	3.9-4.7	6.7-8.0	9.4-11.3
210 - 239	6.8-7.5	11.7-12.3	16.4-18.0	4.7-5.0	8.0-8.5	11.3-12.0
240 - 269	7.5-8.1	12.3-13.9	18.0-19.5	5.0-5.3	8.5-9.0	12.0-12.2
270 - 299	8.1-8.7	13.9-14.8	19.5-20.9	5.3-5.5	9.0-9.4	12.2-13.2
300 - 329	8.7-9.3	14.8-15.8	20.9-22.2	5.5-5.7	9.4-9.7	13.2-13.7
330 - 364	9.3-9.8	15.8-16.8	22.2-23.6	5.7-5.9	9.7-10.1	13.7-14.3
365-	9.8	16.8	23.6	5.9	10.1	14.3
365+	9.8	16.8	23.6	5.9	10.1	14.3

Pm - maximum photosynthesis rate in kg CH_2O ha^{-1} hr^{-1}

TABLE A7.4
Wood biomass yield potential (Bm) without constraints (mean annual increment in t/ha dry weight) at high level of inputs

Length of growing period (days)	Species without nitrogen fixing ability			Species with nitrogen fixing ability		
	Pm=7.5	Pm=15.0	Pm=25.0	Pm=7.5	Pm=15.0	Pm=25.0
1 - 29	0.0-0.2	0.0-0.3	0.0-0.4	0.0-0.1	0.0-0.2	0.0-0.3
30 - 59	0.2-1.2	0.3-2.0	0.4-2.8	0.1-1.0	0.2-1.7	0.3-2.4
60 - 89	1.2-2.5	2.0-4.3	2.8-6.0	1.0-2.0	1.7-3.5	2.4-4.9
90 - 119	2.5-3.7	4.3-6.3	6.0-8.9	2.0-2.9	3.5-4.9	4.9-6.9
120 - 149	3.7-5.4	6.3-9.2	8.9-13.0	2.9-4.0	4.9-6.8	6.9-9.6
150 - 179	5.4-6.6	9.2-11.2	13.0-15.8	4.0-4.7	6.8-8.0	9.6-11.2
180 - 209	6.6-8.2	11.2-14.0	15.8-19.7	4.7-5.6	8.0-9.6	11.2-13.5
210 - 239	8.2-9.0	14.0-15.3	19.7-21.6	5.6-6.0	9.6-10.2	13.5-14.4
240 - 269	9.0-9.7	15.3-16.6	21.6-23.4	6.0-6.3	10.2-10.8	14.4-15.2
270 - 299	9.7-10.4	16.6-17.8	23.4-25.1	6.3-6.6	10.8-11.2	15.2-15.8
300 - 329	10.4-11.1	17.8-18.9	25.1-26.6	6.6-6.8	11.2-11.7	15.8-16.4
330 - 364	11.1-11.8	18.9-20.1	26.6-28.3	6.8-7.1	11.7-12.1	16.4-17.1
365-	11.8	20.1	28.3	7.1	12.1	17.1
365+	11.8	20.1	28.3	7.1	12.1	17.1

Pm - maximum photosynthesis rate in kg CH_2O ha^{-1} hr^{-1}

TABLE A7.5
Wood biomass yield potential (Bm) without constraints (mean annual increment in t/ha dry weight) at intermediate level of inputs

Length of growing period (days)	Species without nitrogen fixing ability			Species with nitrogen fixing ability		
	Pm = 7.5	Pm = 15.0	Pm = 25.0	Pm = 7.5	Pm = 15.0	Pm = 25.0
1 - 29	0.0-0.2	0.0-0.2	0.0-0.3	0.0-0.1	0.0-0.2	0.0-0.2
30 - 59	0.2-0.9	0.2-1.5	0.3-2.1	0.1-0.8	0.2-1.3	0.2-1.8
60 - 89	0.9-1.9	1.5-3.2	2.1-4.5	0.8-1.5	1.3-2.6	1.8-3.7
90 - 119	1.9-2.8	3.2-4.7	4.5-6.7	1.5-2.2	2.6-3.7	3.7-5.2
120 - 149	2.8-4.1	4.7-6.9	6.7-9.8	2.2-3.0	3.7-5.1	5.2-7.2
150 - 179	4.1-5.0	6.9-8.4	9.8-11.9	3.0-3.5	5.1-6.0	7.2-8.4
180 - 209	5.0-6.2	8.4-10.5	11.9-14.8	3.5-4.2	6.0-7.2	8.4-10.1
210 - 239	6.2-6.8	10.5-11.5	14.8-16.2	4.2-4.5	7.2-7.7	10.1-10.8
240 - 269	6.8-7.3	11.5-12.5	16.2-17.6	4.5-4.7	7.7-8.1	10.8-11.4
270 - 299	7.3-7.6	12.5-13.4	17.6-18.8	4.7-5.0	8.1-8.4	11.4-11.9
300 - 329	7.6-8.3	13.4-14.2	18.8-20.0	5.0-5.1	8.4-8.8	11.9-12.3
330 - 364	8.3-8.9	14.2-15.1	20.0-21.2	5.1-5.3	8.8-9.1	12.3-12.8
365-	8.9	15.1	21.2	5.3	9.1	12.8
365+	8.9	15.1	21.2	5.3	9.1	12.8

Pm - maximum photosynthesis rate in kg CH_2O ha^{-1} hr^{-1}

TABLE A7.6
Wood biomass yield potential (Bm) without constraints (mean annual increment in t/ha dry weight) at low level of inputs

Length of growing period (days)	Species without nitrogen fixing ability			Species with nitrogen fixing ability		
	Pm = 7.5	Pm = 15.0	Pm = 25.0	Pm = 7.5	Pm = 15.0	Pm = 25.0
1 - 29	0.0-0.1	0.0-0.2	0.0-0.2	0.0-0.1	0.0-0.1	0.0-0.2
30 - 59	0.1-0.6	0.2-1.0	0.2-1.4	0.1-0.5	0.1-0.9	0.2-1.2
60 - 89	0.6-1.3	1.0-2.2	1.4-3.0	0.5-1.0	0.9-1.8	1.2-2.5
90 - 119	1.3-1.9	2.2-3.2	3.0-4.5	1.0-1.5	1.8-2.5	2.5-3.5
120 - 149	1.9-2.7	3.2-4.6	4.5-6.5	1.5-2.0	2.5-3.4	3.5-4.8
150 - 179	2.7-3.3	4.6-5.6	6.5-7.9	2.0-2.4	3.4-4.0	4.8-5.6
180 - 209	3.3-4.1	5.6-7.0	7.9-9.9	2.4-2.8	4.0-4.8	5.6-6.8
210 - 239	4.1-4.5	7.0-7.7	9.9-10.8	2.8-3.0	4.8-5.1	6.8-7.2
240 - 269	4.5-4.9	7.7-8.3	10.8-11.7	3.0-3.2	5.1-5.4	7.2-7.6
270 - 299	4.9-5.2	8.3-8.9	11.7-12.6	3.2-3.3	5.4-5.6	7.6-7.9
300 - 329	5.2-5.6	8.9-9.5	12.6-13.3	3.3-3.4	5.6-5.9	7.9-8.2
330 - 364	5.6-5.9	9.5-10.1	13.3-14.2	3.4-3.6	5.9-6.1	8.2-8.6
365-	5.9	10.1	14.2	3.6	6.1	8.6
365+	5.9	10.1	14.2	3.6	6.1	8.6

Pm - maximum photosynthesis rate in kg CH_2O ha^{-1} hr^{-1}

WORLD SOIL RESOURCES REPORTS

1. Report of the First Meeting of the Advisory Panel on the Soil Map of the World, Rome, 19-23 June 1961.**
2. Report of the First Meeting on Soil Survey, Correlation and Interpretation for Latin America, Rio de Janeiro, Brazil, 28-31 May 1962**
3. Report of the First Soil Correlation Seminar for Europe, Moscow, USSR, 16-28 July 1962.**
4. Report of the First Soil Correlation Seminar for South and Central Asia, Tashkent, Uzbekistan, USSR, 14 September-2 October 1962.**
5. Report of the Fourth Session of the Working Party on Soil Classification and Survey (Subcommission on Land and Water Use of the European Commission on Agriculture), Lisbon, Portugal, 6-10 March 1963.**
6. Report of the Second Meeting of the Advisory Panel on the Soil Map of the World, Rome, 9-11 July 1963.**
7. Report of the Second Soil Correlation Seminar for Europe, Bucharest, Romania, 29 July-6 August 1963.**
8. Report of the Third Meeting of the Advisory Panel on the Soil Map of the World, Paris, 3 January 1964.**
9. Adequacy of Soil Studies in Paraguay, Bolivia and Peru, November-December 1963.**
10. Report on the Soils of Bolivia, January 1964.**
11. Report on the Soils of Paraguay, January 1964.**
12. Preliminary Definition, Legend and Correlation Table for the Soil Map of the World, Rome, August 1964.**
13. Report of the Fourth Meeting of the Advisory Panel on the Soil Map of the World, Rome, 16-21 May 1964.**
14. Report of the Meeting on the Classification and Correlation of Soils from Volcanic Ash, Tokyo, Japan, 11-27 June 1964.**
15. Report of the First Session of the Working Party on Soil Classification, Survey and Soil Resources of the European Commission on Agriculture, Florence, Italy, 1-3 October 1964.**
16. Detailed Legend for the Third Draft on the Soil Map of South America, June 1965.**
17. Report of the First Meeting on Soil Correlation for North America, Mexico, 1-8 February 1965.**
18. The Soil Resources of Latin America, October 1965.**
19. Report of the Third Correlation Seminar for Europe: Bulgaria, Greece, Romania, Turkey, Yugoslavia, 29 August-22 September 1965.**
20. Report of the Meeting of Rapporteurs, Soil Map of Europe (Scale 1:1 000 000) (Working Party on Soil Classification and Survey of the European Commission on Agriculture), Bonn, Federal Republic of Germany, 29 November-3 December 1965.**
21. Report of the Second Meeting on Soil Survey, Correlation and Interpretation for Latin America, Rio de Janeiro, Brazil, 13-16 July 1965.**
22. Report of the Soil Resources Expedition in Western and Central Brazil, 24 June-9 July 1965.**
23. Bibliography on Soils and Related Sciences for Latin America (1st edition), December 1965.**
24. Report on the Soils of Paraguay (2nd edition), August 1964.**
25. Report of the Soil Correlation Study Tour in Uruguay, Brazil and Argentina, June-August 1964.**
26. Report of the Meeting on Soil Correlation and Soil Resources Appraisal in India, New Delhi, India, 5-15 April 1965.**
27. Report of the Sixth Session of the Working Party on Soil Classification and Survey of the European Commission on Agriculture, Montpellier, France, 7-11 March 1967.**
28. Report of the Second Meeting on Soil Correlation for North America, Winnipeg-Vancouver, Canada, 25 July-5 August 1966.**
29. Report of the Fifth Meeting of the Advisory Panel on the Soil Map of the World, Moscow, USSR, 20-28 August 1966.**
30. Report of the Meeting of the Soil Correlation Committee for South America, Buenos Aires, Argentina, 12-19 December 1966.**
31. Trace Element Problems in Relation to Soil Units in Europe (Working Party on Soil Classification and Survey of the European Commission on Agriculture), Rome, 1967.**
32. Approaches to Soil Classification, 1968.**
33. Definitions of Soil Units for the Soil Map of the World, April 1968.**
34. Soil Map of South America 1:5 000 000, Draft Explanatory Text, November 1968.**
35. Report of a Soil Correlation Study Tour in Sweden and Poland, 27 September-14 October 1968.**
36. Meeting of Rapporteurs, Soil Map of Europe (Scale 1:1 000 000) (Working Party on Soil Classification and Survey of the European Commission on Agriculture), Poitiers, France 21-23 June 1967.**
37. Supplement to Definition of Soil Units for the Soil Map of the World, July 1969.**
38. Seventh Session of the Working Party on Soil Classification and Survey of the European Commission on Agriculture, Varna, Bulgaria, 11-13 September 1969.**
39. A Correlation Study of Red and Yellow Soils in Areas with a Mediterranean Climate.**
40. Report of the Regional Seminar of the Evaluation of Soil Resources in West Africa, Kumasi, Ghana, 14-19 December 1970.**
41. Soil Survey and Soil Fertility Research in Asia and the Far East, New Delhi, 15-20 February 1971.**
42. Report of the Eighth Session of the Working Party on Soil Classification and Survey of the European Commission on Agriculture, Helsinki, Finland, 5-7 July 1971.**
43. Report of the Ninth Session of the Working Party on Soil Classification and Survey of the European Commission on

Agriculture, Ghent, Belgium 28-31 August 1973.**

44. First Meeting of the West African Sub-Committee on Soil Correlation for Soil Evaluation and Management, Accra, Ghana, 12-19 June 1972.**
45. Report of the Ad Hoc Expert Consultation on Land Evaluation, Rome, Italy, 6-8 January 1975.**
46. First Meeting of the Eastern African Sub-Committee for Soil Correlation and Land Evaluation, Nairobi, Kenya, 11-16 March 1974.**
47. Second Meeting of the Eastern African Sub-Committee for Soil Correlation and Land Evaluation, Addis Ababa, Ethiopia, 25-30 October 1976.
48. Report on the Agro-Ecological Zones Project, Vol. 1 - Methodology and Results for Africa, 1978. Vol. 2 - Results for Southwest Asia, 1978.
49. Report of an Expert Consultation on Land Evaluation Standards for Rainfed Agriculture, Rome, Italy, 25-28 October 1977.
50. Report of an Expert Consultation on Land Evaluation Criteria for Irrigation, Rome, Italy, 27 February-2 March 1979.
51. Third Meeting of the Eastern African Sub-Committee for Soil Correlation and Land Evaluation, Lusaka, Zambia, 18-30 April 1978.
52. Land Evaluation Guidelines for Rainfed Agriculture, Report of an Expert Consultation, 12-14 December 1979.
53. Fourth Meeting of the West African Sub-Committee for Soil Correlation and Land Evaluation, Banjul, The Gambia, 20-27 October 1979.
54. Fourth Meeting of the Eastern African Sub-Committee for Soil Correlation and Land Evaluation, Arusha, Tanzania, 27 October-4 November 1980.
55. Cinquième réunion du Sous-Comité Ouest et Centre africain de corrélation des sols pour la mise en valeur des terres, Lomé, Togo, 7-12 décembre 1981.
56. Fifth Meeting of the Eastern African Sub-Committee for Soil Correlation and Land Evaluation, Wad Medani, Sudan, 5-10 December 1983.
57. Sixième réunion du Sous-Comité Ouest et Centre Africain de corrélation des sols pour la mise en valeur des terres, Niamey, Niger, 6-12 février 1984.
58. Sixth Meeting of the Eastern African Sub-Committee for Soil Correlation and Land Evaluation, Maseru, Lesotho, 9-18 October 1985.
59. Septième réunion du Sous-Comité Ouest et Centre africain de corrélation des sols pour la mise en valeur des terres, Ouagadougou, Burkina Faso, 10-17 novembre 1985.
60. Revised Legend, Soil Map of the World, FAO-Unesco-ISRIC, 1988. Reprinted 1990.
61. Huitième réunion du Sous-Comité Ouest et Centre africain de corrélation des sols pour la mise en valeur des terres, Yaoundé, Cameroun, 19-28 janvier 1987.
62. Seventh Meeting of the East and Southern African Sub-Committee for Soil Correlation and Evaluation, Gaborone, Botswana, 30 March-8 April 1987.
63. Neuvième réunion du Sous-Comité Ouest et Centre africain de corrélation des sols pour la mise en valeur des terres, Cotonou, Bénin, 14-23 novembre 1988.
64. FAO-ISRIC Soil Database (SDB), 1989.
65. Eighth Meeting of the East and Southern African Sub-Committee for Soil Correlation and Land Evaluation, Harare, Zimbabwe, 9-13 October 1989.
66. World soil resources. An explanatory note on the FAO World Soil Resources Map at 1:25 000 000 scale, 1991.
67. Digitized Soil Map of the World, Volume 1: Africa, Release 1.0, November 1991. Volume 2: North and Central America. Volume 3: Central and South America. Volume 4: Europe and West of the Urals. Volume 5: North East Asia. Volume 6: Near East and Far East. Volume 7: South East Asia and Oceanic.
68. Land Use Planning Applications. Proceedings of the FAO Expert Consultation 1990, Rome, 10-14 December 1990.
69. Dixième réunion du Sous-Comité Ouest et Centre africain de corrélation des sols pour la mise en valeur des terres, Bouaké, Odienné, Côte d'Ivoire, Côte d'Ivoire, 5-12 november 1990.
70. Ninth Meeting of the East and Southern African Sub-Committee for Soil Correlation and Land Evaluation, Lilongwe, Malawi, 25 November - 2 December 1991.
71. Agro-ecological land resources assessment for agricultural development planning. A case study of Kenya. Resources data base and land productivity. Main Report. Technical Annex 1: Land resources. Technical Annex 2: Soil erosion and productivity. Technical Annex 3: Agro-climatic and agro-edaphic suitabilities for barley, oat, cowpea, green gram and pigeonpea. Technical Annex 4: Crop productivity. Technical Annex 5: Livestock productivity. Technical Annex 6: Fuelwood productivity. Technical Annex 7: Systems documentation guide to computer programs for land productivity assessments. Technical Annex 8: Crop productivity assessment: results at district level. 1991.

** Out of print